ARDUINO IR REM

Arduino IR Remote Control, LED Scroll Bar, Digital Clock with Alarm, ATmega328 Chip, Servo Motor Control etc,...

Arduino IR Remote Control, LED Scroll Bar, Digital Clock with Alarm,

ATmega328 Chip, Servo Motor Control etc,...

CONTENTS

Acknowledgments 5

Introduction 6

1. Beginning with Arduino Due 7

2. DIY Arduino LED Scroll Bar 16

3. Arduino Based Digital Clock with Alarm 35

4. Recurrence Counter utilizing Arduino 48

5. Interfacing Arduino with Raspberry Pi utilizing Serial Communication 58

6. Make Your Own Custom made Arduino Board with ATmega328 Chip 67

7. Interfacing TFT LCD with Arduino 78

8. Servo Motor Control with Arduino Due 86

9. PWM with Arduino Due 94

10. Widespread IR Remote Control utilizing Arduino and Android App 102

Thank you !!! 125

ACKNOWLEDGMENTS

The writer might want to recognize the diligent work of the article group in assembling this book. He might likewise want to recognize the diligent work of the Raspberry Pi Foundation and the Arduino bunch for assembling items and networks that help to make the Internet of Things increasingly open to the overall population. Yahoo for the democratization of innovation!

INTRODUCTION

The Internet of Things (IOT) is a perplexing idea comprised of numerous PCs and numerous correspondence ways. Some IOT gadgets are associated with the Internet and some are most certainly not. Some IOT gadgets structure swarms that convey among themselves. Some are intended for a solitary reason, while some are increasingly universally useful PCs. This book is intended to demonstrate to you the IOT from the back to front. By structure IOT gadgets, the per user will comprehend the essential ideas and will almost certainly develop utilizing the rudiments to make his or her very own IOT applications. These included ventures will tell the per user the best way to assemble their very own IOT ventures and to develop the models appeared. The significance of Computer Security in IOT gadgets is additionally talked about and different systems for protecting the IOT from unapproved clients or programmers. The most significant takeaway from this book is in structure the tasks yourself.

1. BEGINNING WITH ARDUINO DUE

ARDUINO DUE is an ARM controller based board intended for electronic Engineers as well as Hobbyists. ARM engineering is extremely compelling in present day gadgets. We are utilizing the ARM design based controllers all over the place. For instance we are utilizing ARM controllers in our mobiles, iPods as well as PCs and so forth. In the event that somebody needs to plan modern frameworks, it must be on ARM controllers. ARM controllers are significant in light of the recurrence of their activity and information transport size.

ARM controllers can accomplish results superior to anything ordinary controllers and they have a greater number of capacities than a typical controller. With this, clearly we should learn ARM controller for planning higher capacities like picture preparing and so on.

To comprehend the ARM engineering, most ideal approach to do is by considering the ARDUINO DUE. Beneath figure shows Arduino Due board.

There are various kinds of ARDUINO sheets in the market, with UNO being the most famous and DUE being the most refined. DUE center is from "SAM3X8E" controller as appeared in figure. This controller works at 84 MHz clock, which is in excess of multiple times the speed of UNO. With just about 60 GPIO (General reason Input Output) we can utilize this board freely, with no need of move registers. We have just secured numerous Arduino and Arduino Uno Projects, from learner to cutting edge level and they spread practically all subjects to take in Arduino without any preparation.

UNO structured from ATMEGA controller, which is 8 piece type, and DUE planned from ARM type, which is a 32 piece type. This number itself separates the accomplishment, power and speed hole between two sheets. We picked DUE board since it is the most effortless approach to comprehend ARM controller particularly first off. So in this instructional exercise we are gonna to Blink a LED utilizing Arduino Due, for beginning with Arduino Due Board. This Program and instructional exercise will likewise goes with Arduino Uno to flicker LED with it. The product and download, transfer procedure are same for the Uno.

The ARDUINO DUE sheets additionally have Shield sheets, they are fundamentally augmentations for ARDUINO. These shields add extra highlights to the ARDUINO. These shields are stacked one over the other on ARDUINO.

Required Components:

Equipment: Arduino Due board, interfacing pins, 220? resistor, LED, bread board.

Programming: Arduino daily, download it from this connection: https://www.arduino.cc/en/Main/Software

Open the above given connection, under download session, we have the most recent adaptation of ARDUINO programming, which is 1.6.8 (at the hour of composing this article). Regardless of whether you have the more seasoned variant, download the more up to date form. In past adaptations the DUE board libraries are absent. So the past renditions can't identify the DUE board. You can refresh the past variant to get the DUE board working.

Snap on the windows installer button for the product:

Download the Arduino Software

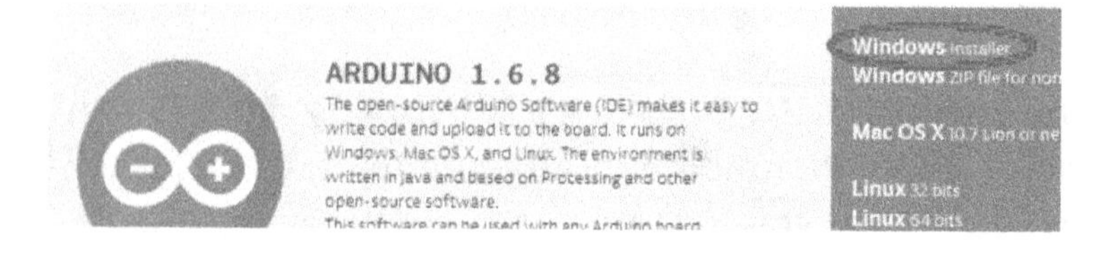

Presently click on the simply download button for the arrangement to begin downloading. The arrangement document would be around 85 Mb.

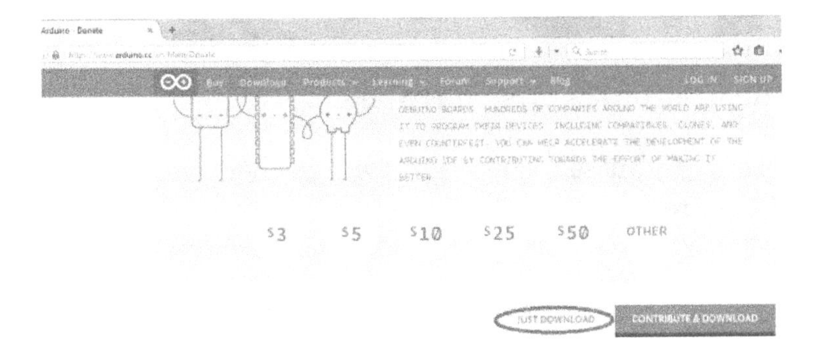

After download introduce the document by double tap. Once the introduce is finished, you will get a symbol on the work area as demonstrated as follows.

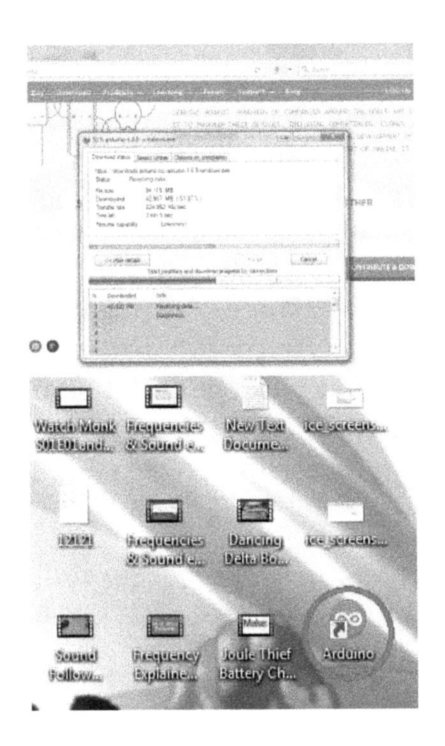

Double tap on the program to begin.

Presently you see, there are two connectives on the DUE board.

Both of the ports can be utilized to program the DUE, yet we are going to utilize NATIVE USB port. Presently interface the USB plug and associate the opposite end to PC, you should notice the power LED ON.

When the ARDUINO program is running, you ought to pick the ARDUINO DUE board from the 'Instruments' menu of the program. When you pick the DUE board you will see the chose board at the correct base, as appeared in the figure given underneath in next segment.

We associated the USB to NATIVE port, so we need to pick the 'Local port' in the product. This choice will likewise be in 'Devices' choice. When you select it, you are prepared to transfer the program.

Circuit and Working Explanation:

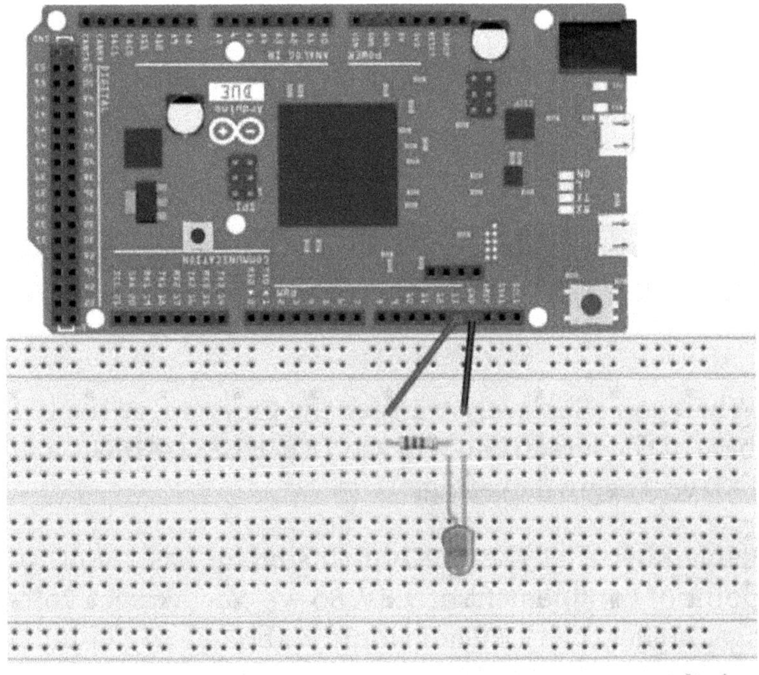

In here we will compose a program to squint a LED for each 1000ms.We will associate a LED at PIN13 through a 220? current restricting resistor.

Presently transfer the program by tapping on the Upload Button, appeared in the figure (upper left corner),

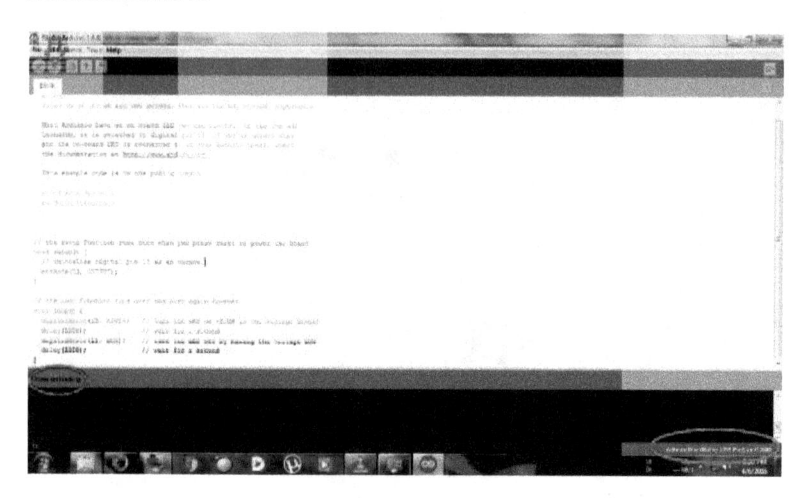

When you effectively transfer the program, at the left base of the screen you will see 'DONE UPLOADING' and LED will begin flickering.

Remember that the GPIO of this board has a voltage point of confinement of 3.3V. So we can't expect voltages higher than 3.3V nor would we be able to give voltages higher than 3.3V to any stick of this board. In the event that voltage higher than 3.3v is given to board, at that point it could harm the board for all time.

Check the Code underneath to show signs of improvement understanding.

Code

```
void setup()
{
 //initialize digital pin 13 as an output.
```

```
pinMode(13,OUTPUT);
}
//the loop function runs over and over again forever
void loop()
{

digitalWrite(13,HIGH); //turn the LED on (making the voltage level HIGH)

delay(1000);        //wait for a second

digitalWrite(13, LOW);    // turn the LED off (making the voltage level LOW)

delay(1000);        //wait for a second
}
```

◆ ◆ ◆

2. DIY ARDUINO LED SCROLL BAR

I am gonna to share a LED Scroll Bar as appeared in the above picture. Ten LED strips can streak in various impacts by utilizing an Arduino board.

Required Components:

- Arduino Nano
- LED strip
- Control board
- Dupont line

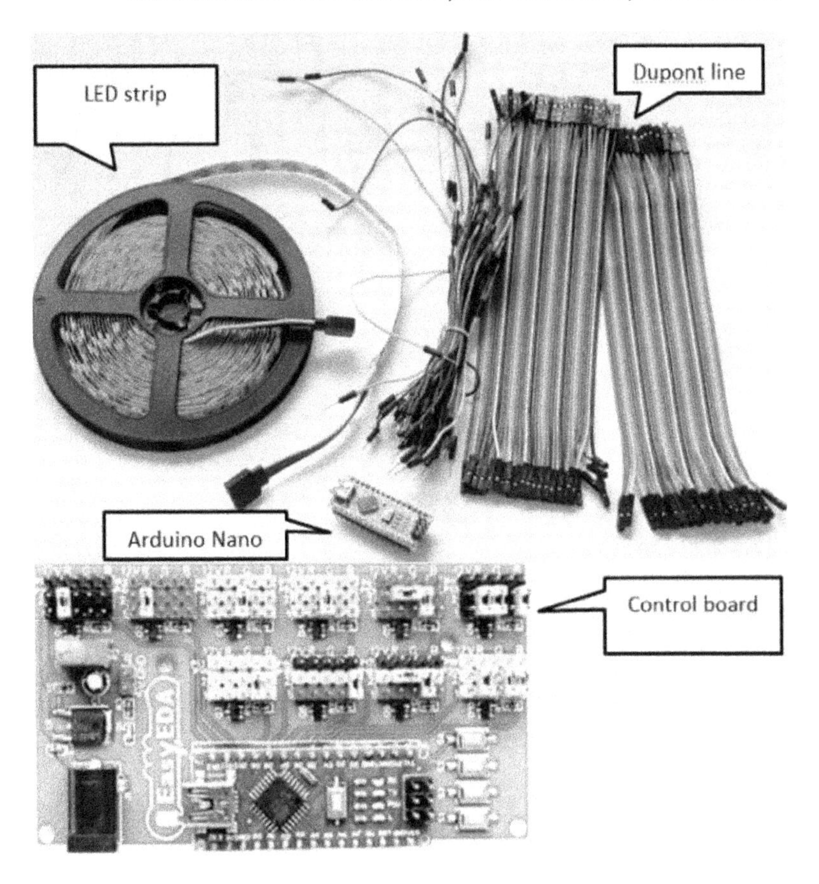

Steps for making the LED Scroll Bar:

Stage 1) Preparation

Cut the LED strip into ten pieces as well as each piece has discretionary number of LEDs.

At that point, weld the navigate at an interface of the LED strip. Here I want to utilize Dupont line to interface.

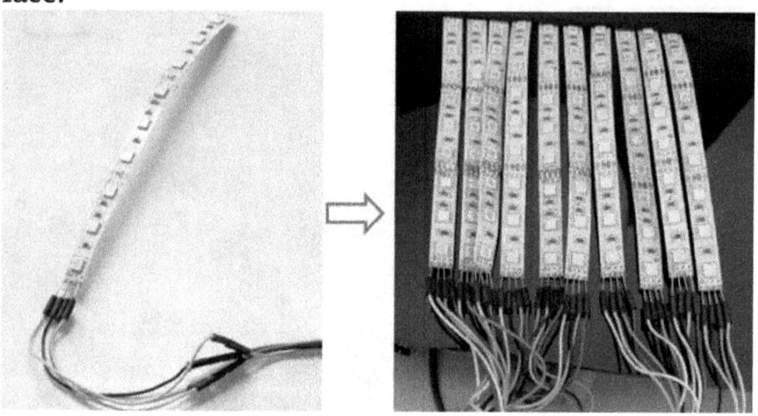

Stage 2) Design

Make a control board. Control board is utilizing to control the glimmer example of LED strips. Helped by Arduino Nano and outfitted with a keypad, we can make the LED strip streak in various examples.

Step 2.1) Start with Designing Schematic

To structure my circuit, I pick a free online EDA apparatus called EasyEDA which is a one stop configuration look for your hardware ventures. It offers schematic catch, zest reproduction, PCB configuration for nothing and furthermore offers high caliber however low value Customized PCB administration. There are an enormous number of segment libraries in its manager, so you can without quite a bit of a stretch and rapidly locate your ideal parts. Check here the total instructional exercise on How to utilize Easy EDA for making

Schematics, PCB formats, Simulating the Circuits and so forth.

You can get to the Schematic outline of this drove parchment bar by following this connection.

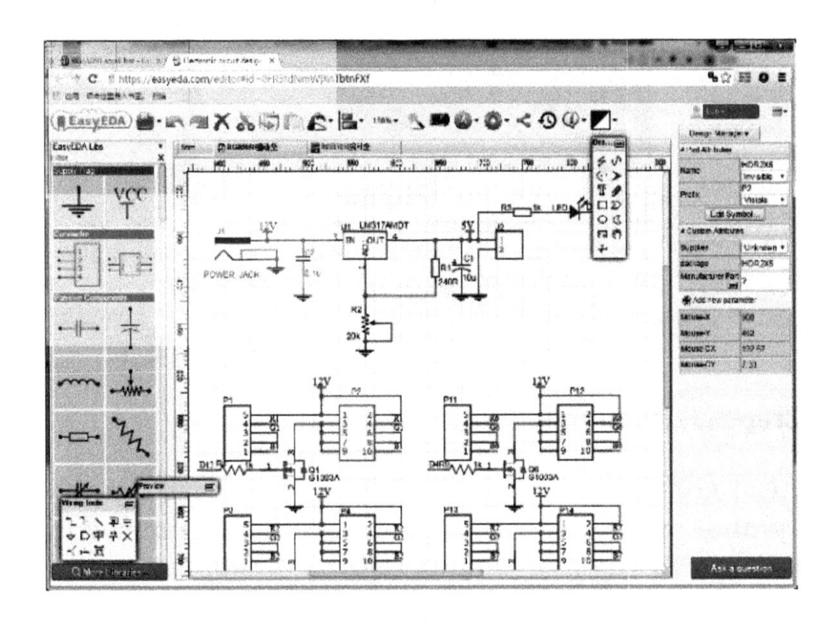

Notes: The voltage of the LED strip is 12 V as well as Arduino Nano is 5V. Make sure to include a power controller, for example, AMS1117-5.0.

Step 2.2) Create the PCB Layout.

You will notice the PCB design in the accompanying graph:

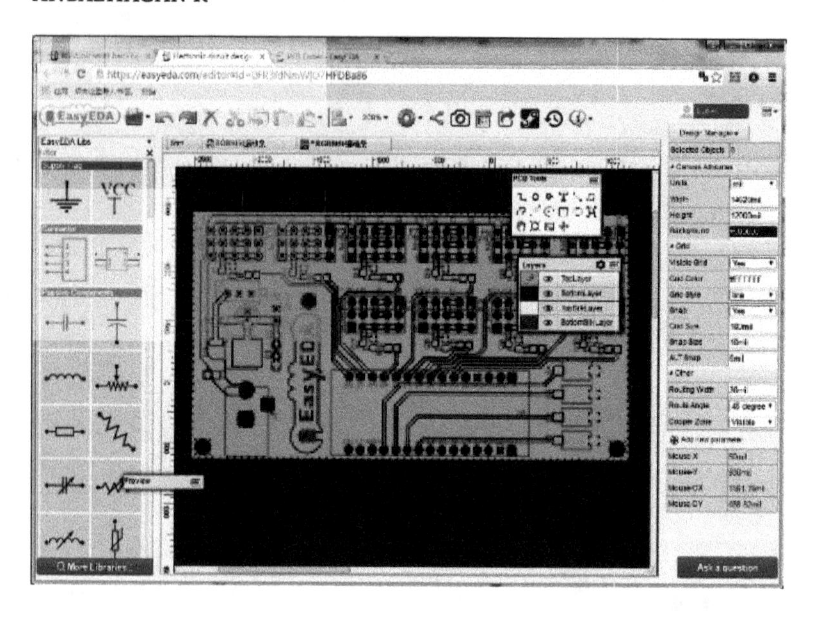

Step 2.3) Make an example

After complete the structure of PCB, you can tap the symbol of Fabrication yield above. At that point you will get to the page PCB request to download Gerber records of your PCB and send them to any producer, it's likewise significantly simpler (and less expensive) to arrange it straightforwardly in EasyEDA. Here you can choose the quantity of PCBs you need to arrange, what number of copper layers you need, the PCB thickness, copper weight, and even the PCB shading. After you have chosen the entirety of the choices, click "Spare to Cart" and complete you request, at that point you will get your PCBs a couple of days after the fact.

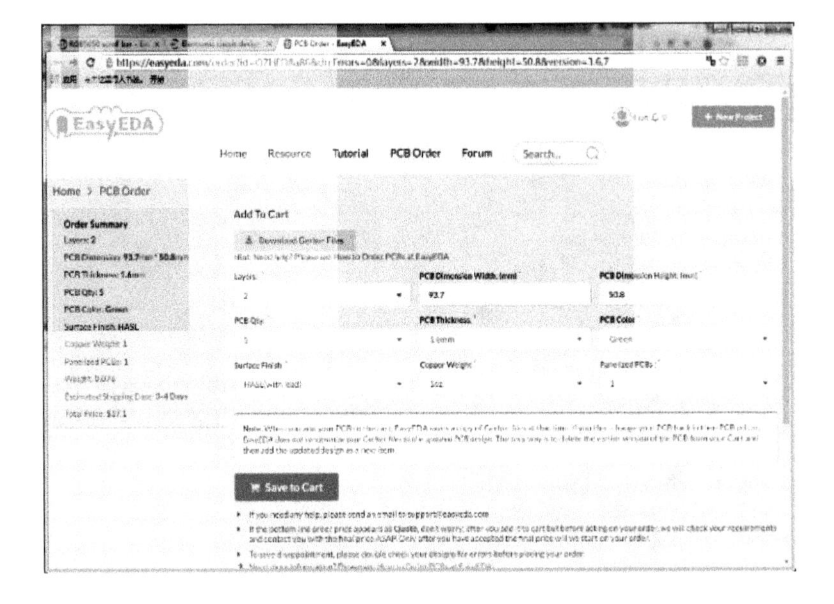

Step 2.4) Take conveyance of the PCB

At the point when I got the PCBs, I am very intrigued with the quality, they are truly pleasant.

Step 2.5) Welding

It is exceptionally simple to make a control board. Similarly as the accompanying picture depicted, when the parts are welded, it is finished.

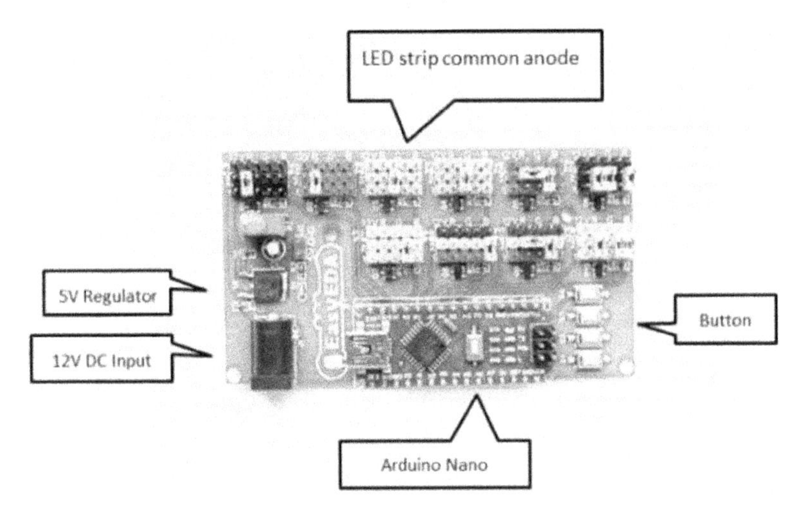

Stage 3) Connection

Associate the LED strip to the control load up and simultaneously please focus on the positive and negative of the terminals.

As the image appeared beneath.

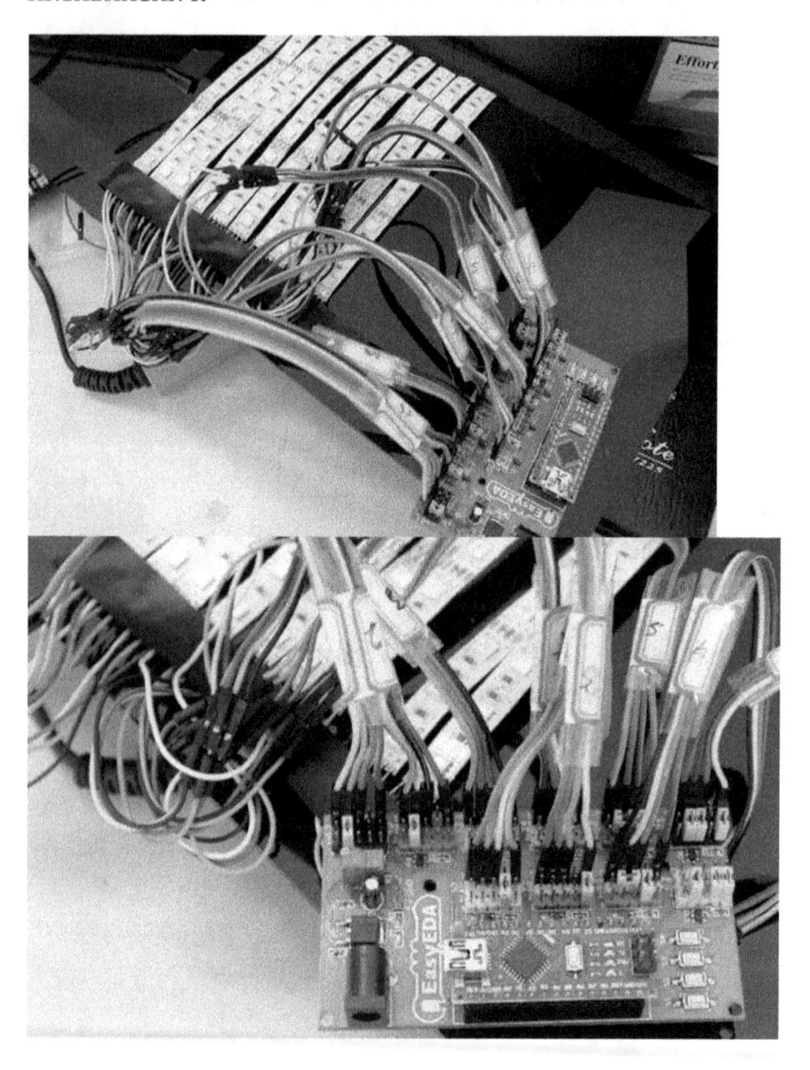

Stage 4) Download a program

Associate it to a 12 V power supply, download a program (Check the full code beneath) on the Arduino Nano and run it.

Press the catch to switch streak mode.

On the off chance that you need, you can clone my LED Scroll Bar Schematic and PCB here. Additionally you can get to the Arduino Code, Required Components and different subtleties of this LED Scroll bar by following the given connection.

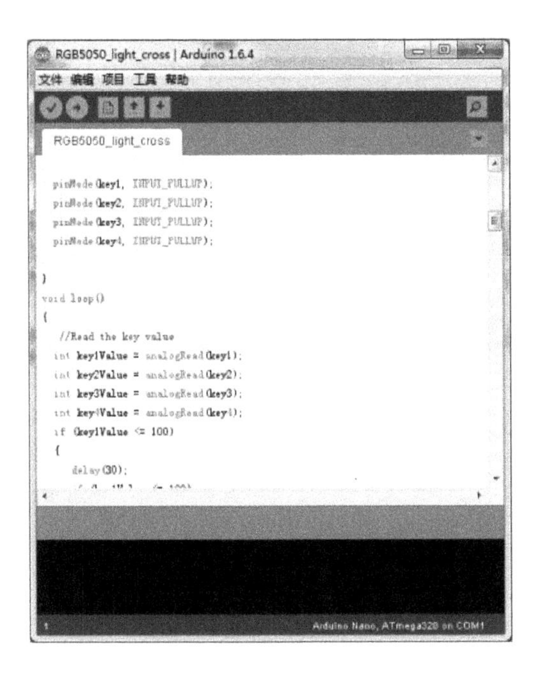

Presently I finished my undertaking of parchment bar.

Making your own LED Scroll Bar makes a great deal of fun, and the outcomes can be fulfilling. Ideally this article will help you with making a LED Scroll Bar, You can likewise compose a program to make the LED strip streak in the manner you like.

Code

```
int RGB1 =12;
int RGB2 =11;
int RGB3 =10;
int RGB4 =9;
int RGB5 =8;
int RGB6 =7;
int RGB7 =6;
int RGB8 =5;
int RGB9 =4;
int RGB10 =3;
int key1 =A3;
```

```
int key2 =A2;
int key3 =A1;
int key4 =A0;
void setup()
{
 Serial.begin(9600);


  pinMode(RGB1, OUTPUT);
 pinMode(RGB2, OUTPUT);
 pinMode(RGB3, OUTPUT);
 pinMode(RGB4, OUTPUT);
 pinMode(RGB5, OUTPUT);
 pinMode(RGB6, OUTPUT);
 pinMode(RGB7, OUTPUT);
 pinMode(RGB8, OUTPUT);
 pinMode(RGB9, OUTPUT);
 pinMode(RGB10, OUTPUT);


  pinMode(key1, INPUT_PULLUP);
 pinMode(key2, INPUT_PULLUP);
 pinMode(key3, INPUT_PULLUP);
 pinMode(key4, INPUT_PULLUP);


}
void loop()
{
 int key1Value = analogRead(key1);
 int key2Value = analogRead(key2);
```

```
int key3Value = analogRead(key3);
int key4Value = analogRead(key4);
if(key1Value <= 100)
{
 delay(30);
 if(key1Value <= 100)
 {
  digitalWrite(RGB5,HIGH);
  digitalWrite(RGB6, HIGH);
  delay(50);
  digitalWrite(RGB4, HIGH);
  digitalWrite(RGB7,HIGH);
  delay(50);
  digitalWrite(RGB3, HIGH);
  digitalWrite(RGB8, HIGH);
  delay(50);
  digitalWrite(RGB2,HIGH);
  digitalWrite(RGB9, HIGH);
  delay(50);
  digitalWrite(RGB1, HIGH);
  digitalWrite(RGB10, HIGH);
  delay(1000);
 }
 else
 {
  digitalWrite(RGB1, LOW);
  digitalWrite(RGB2, LOW);
  digitalWrite(RGB3, LOW);
  digitalWrite(RGB4, LOW);
  digitalWrite(RGB5, LOW);
```

```
    digitalWrite(RGB6, LOW);
    digitalWrite(RGB7, LOW);
    digitalWrite(RGB8, LOW);
    digitalWrite(RGB9, LOW);
    digitalWrite(RGB10, LOW);
   }
 }


  if(key2Value <= 100)
  {
   digitalWrite(RGB1,HIGH);
   digitalWrite(RGB6, HIGH);
   delay(40);
   digitalWrite(RGB2, HIGH);
   digitalWrite(RGB7,HIGH);
   delay(40);
   digitalWrite(RGB3, HIGH);
   digitalWrite(RGB8, HIGH);
   delay(40);
   digitalWrite(RGB4,HIGH);
   digitalWrite(RGB9, HIGH);
   delay(40);
   digitalWrite(RGB5, HIGH);
   digitalWrite(RGB10, HIGH);
   delay(1000);
  }
  else
  {
   digitalWrite(RGB1, LOW);
   digitalWrite(RGB2, LOW);
```

```
  digitalWrite(RGB3, LOW);
  digitalWrite(RGB4, LOW);
  digitalWrite(RGB5, LOW);
  digitalWrite(RGB6, LOW);
  digitalWrite(RGB7, LOW);
  digitalWrite(RGB8, LOW);
  digitalWrite(RGB9, LOW);
  digitalWrite(RGB10, LOW);
}
if(key3Value <= 100)
{
  digitalWrite(RGB1,HIGH);
  delay(90);
  digitalWrite(RGB1, LOW);
  digitalWrite(RGB2, HIGH);
  delay(90);
  digitalWrite(RGB2, LOW);
  digitalWrite(RGB3, HIGH);
  delay(90);
  digitalWrite(RGB3, LOW);
  digitalWrite(RGB4, HIGH);
  delay(90);
  digitalWrite(RGB4, LOW);
  digitalWrite(RGB5,HIGH);
  delay(90);
  digitalWrite(RGB5, LOW);
  digitalWrite(RGB6,HIGH);
  delay(90);
  digitalWrite(RGB6, LOW);
  digitalWrite(RGB7,HIGH);
```

```
 delay(90);
 digitalWrite(RGB7, LOW);
 digitalWrite(RGB8,HIGH);
 delay(90);
 digitalWrite(RGB8, LOW);
 digitalWrite(RGB9,HIGH);
 delay(90);
 digitalWrite(RGB9, LOW);
 digitalWrite(RGB10,HIGH);
 delay(1000);
}
else
{
 digitalWrite(RGB1, LOW);
 digitalWrite(RGB2, LOW);
 digitalWrite(RGB3, LOW);
 digitalWrite(RGB4, LOW);
 digitalWrite(RGB5, LOW);
 digitalWrite(RGB6, LOW);
 digitalWrite(RGB7, LOW);
 digitalWrite(RGB8, LOW);
 digitalWrite(RGB9, LOW);
 digitalWrite(RGB10, LOW);
}
if(key4Value <= 100)
{
 digitalWrite(RGB1,HIGH);
 delay(50);
 digitalWrite(RGB1, LOW);
 digitalWrite(RGB2, HIGH);
```

```
delay(50);
digitalWrite(RGB2, LOW);
digitalWrite(RGB3, HIGH);
delay(50);
digitalWrite(RGB3, LOW);
digitalWrite(RGB4,HIGH);
delay(50);
digitalWrite(RGB4, LOW);
digitalWrite(RGB5, HIGH);
delay(50);
digitalWrite(RGB5, LOW);
digitalWrite(RGB6, HIGH);
delay(50);
digitalWrite(RGB6, LOW);
digitalWrite(RGB7,HIGH);
delay(50);
digitalWrite(RGB7, LOW);
digitalWrite(RGB8, HIGH);
delay(50);
digitalWrite(RGB8, LOW);
digitalWrite(RGB9, HIGH);
delay(50);
digitalWrite(RGB9, LOW);
digitalWrite(RGB10,HIGH);
delay(50);
digitalWrite(RGB10, LOW);
digitalWrite(RGB9, HIGH);
delay(50);
digitalWrite(RGB9, LOW);
digitalWrite(RGB8, HIGH);
```

```
 delay(50);
 digitalWrite(RGB8, LOW);
 digitalWrite(RGB7,HIGH);
 delay(50);
 digitalWrite(RGB7, LOW);
 digitalWrite(RGB6, HIGH);
 delay(50);
 digitalWrite(RGB6, LOW);
 digitalWrite(RGB5, HIGH);
 delay(50);
 digitalWrite(RGB5, LOW);
 digitalWrite(RGB4,HIGH);
 delay(50);
 digitalWrite(RGB4, LOW);
 digitalWrite(RGB3, HIGH);
 delay(50);
 digitalWrite(RGB3, LOW);
 digitalWrite(RGB2, HIGH);
 delay(50);
 digitalWrite(RGB2, LOW);
 digitalWrite(RGB1, HIGH);
 delay(50);
 digitalWrite(RGB1, LOW);
 delay(1000);
}
else
{
 digitalWrite(RGB1, LOW);
 digitalWrite(RGB2, LOW);
 digitalWrite(RGB3, LOW);
```

```
  digitalWrite(RGB4, LOW);
  digitalWrite(RGB5, LOW);
  digitalWrite(RGB6, LOW);
  digitalWrite(RGB7, LOW);
  digitalWrite(RGB8, LOW);
  digitalWrite(RGB9, LOW);
  digitalWrite(RGB10, LOW);
 }


}
```

◆◆◆

3. ARDUINO BASED DIGITAL CLOCK WITH ALARM

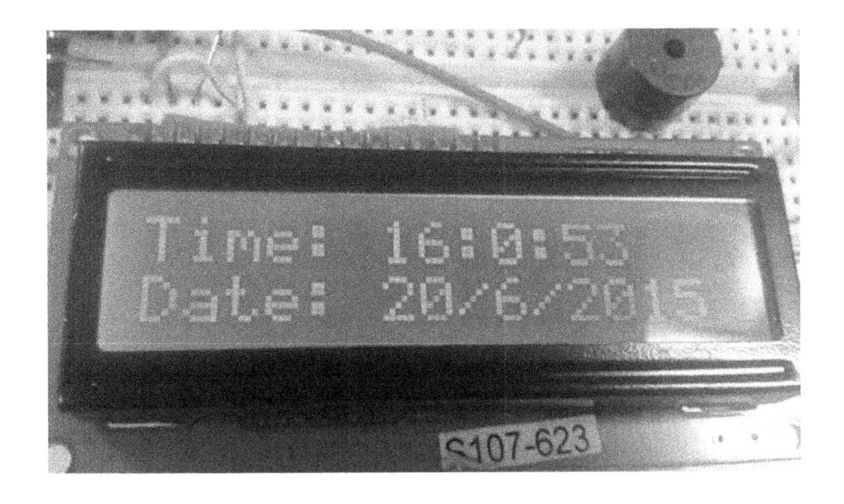

This Arduino based Real time clock is an advanced clock to show continuous utilizing a RTC IC DS1307 which chips away at I2C convention. Constant clock implies it pursues even control disappointment. At the point when power is reconnected, it shows the constant irespective to the time and span it was in off state. In this venture we have utilized a 16x2 LCD module to show the time in - (hour, minute, seconds, date, month and year) group. An Alarm alternative is likewise included and we can set up the alert time. When alert time it spared in inward EEPROM of arduino, it stays spared considerably after reset or power disappointment. Constant checks are generally utilized in our PCs, houses, workplaces and gadgets gadget for keeping them refreshed with ongoing.

I2C convention is a strategy to associate at least two gadgets utilizing two wires to a solitary framework, thus this convention is likewise called as two wire convention. It tends to be utilized to impart 127 gadgets to a solitary gadget or processor. The greater part of I2C gadgets run on 100Khz recurrence.

Steps for information composing expert to (slave getting mode)

- Sends START condition to slave.
- Sends slave address to slave.
- Send compose bit (0) to slave.
- Gotten ACK bit from slave
- Sends words address to slave.
- Gotten ACK bit from slave
- Sends information to slave.
- Gotten ACK bit from slave.
- What's more, last sends STOP condition to

slave.

Ventures for information perusing from slave to ace (slave transmitting mode)

- Sends START condition to slave.
- Sends slave address to slave.
- Send read bit (1) to slave.
- Gotten ACK bit from slave
- Gotten information from slave
- Gotten ACK bit from slave.
- Sends STOP condition to slave.

Circuit Diagram and Description

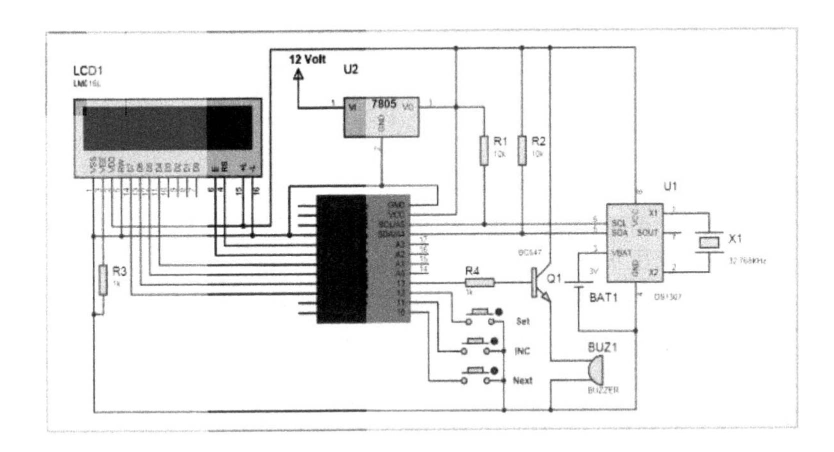

In this Arduino based advanced clock circuit, we have utilized three significant segments which are IC DS1307, Arduino Pro Mini Board and 16x2 LCD module.

Here arduino is utilized for perusing time from ds1307 and show it on 16x2 LCD. DS1307 sends time/date utilizing 2 lines to arduino. A ringer is additionally utilized for caution sign, which signals when alert is enacted. A square outline is appeared beneath to comprehend the working of this Real Time Clock.

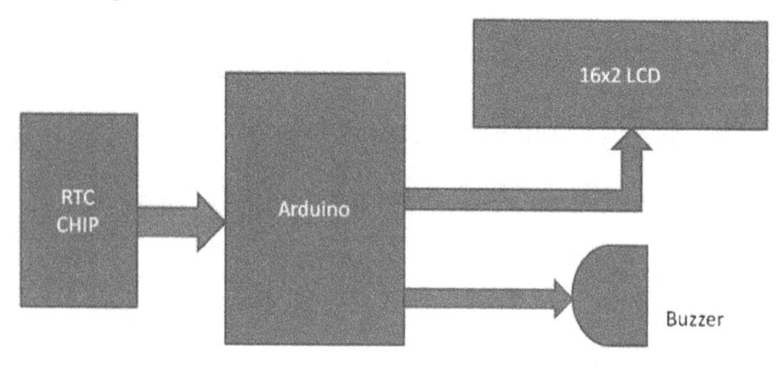

As should be obvious in the circuit chart, DS1307 chip stick SDA and SCL are associated with arduino pins SDA and SCL with draw up resistor that holds default esteem HIGH at information as well as clock lines. 32.768KHz precious stone oscillator is associated with DS1307chip for creating definite 1 second delay, and a 3 volt battery is likewise associated with stick third (BAT) of DS1307 which keeps time pursuing power failure.16x2 LCD is associated with arduino in 4-piece mode. Control stick RS, RW and En are legitimately associated with arduino stick 2, GND and 3. Furthermore, information stick D0-D7 is associated with 4, 5, 6, 7 of arduino. A ringer is associated with arduino stick number 13 through a NPN BC547 transistor having a 1 k resistor at its base.

Three fastens specifically set, INC and Next are utilized for setting caution to stick 12, 11 and 10 of arduino in dynamic low mode. At the point when we press set, caution set mode enacts and now we have to set alert by utilizing INC button and Next catch is utilized for moving to digit. The total breadboard arrangement of this continuous clock with caution is appeared in underneath picture. You can likewise check a point by point instructional exercise on advanced morning timer with AVR microcontroller.

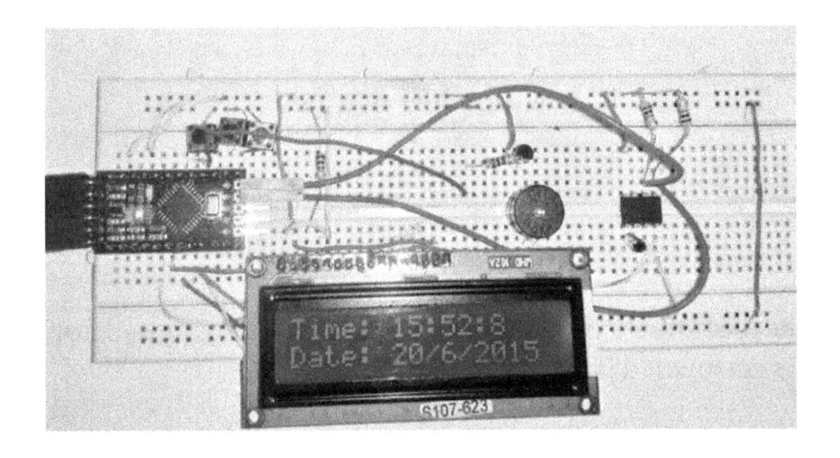

Program Description

To program for this constant clock, we have utilized a few libraries for removing time/date from DS1307 and for showing on LCD, which are given beneath:

```
#include <Wire.h>
#include<EEPROM.h>
#include <RTClib.h>
#include <LiquidCrystal.h>
```

Also, instatement of RTC, LCD and info yield are performed in arrangement circle.

```
void setup()
{
 Wire.begin();
 RTC.begin();
 lcd.begin(16,2);
 pinMode(INC, INPUT);
 pinMode(next, INPUT);
 pinMode(set_mad, INPUT);
 pinMode(buzzer, OUTPUT);
 digitalWrite(next, HIGH);
 digitalWrite(set_mad, HIGH);
 digitalWrite(INC, HIGH);
```

Rest of things like understanding time, setting caution is performed in void circle area.

```
lcd.print("Time:");
lcd.setCursor(6,0);
lcd.print(HOUR=now.hour(),DEC);
lcd.print(":");
lcd.print(MINUT=now.minute(),DEC);
lcd.print(":");
lcd.print(SECOND=now.second(),DEC);
```

Code

```
/* ----- C Program for Arduino based Alarm Clock ---- */
#include <Wire.h>
#include<EEPROM.h>
#include <RTClib.h>
#include <LiquidCrystal.h>
LiquidCrystal lcd(3, 2, 4, 5, 6, 7);
RTC_DS1307 RTC;
int temp,inc,hours1,minut,add=11;
int next=10;
int INC=11;
int set_mad=12;
#define buzzer 13
int HOUR,MINUT,SECOND;


void setup()
{
 Wire.begin();
 RTC.begin();
 lcd.begin(16,2);
 pinMode(INC, INPUT);
 pinMode(next, INPUT);
```

```
pinMode(set_mad, INPUT);
pinMode(buzzer, OUTPUT);
digitalWrite(next, HIGH);
digitalWrite(set_mad, HIGH);
digitalWrite(INC, HIGH);


  lcd.setCursor(0,0);
 lcd.print("Real Time Clock");
 lcd.setCursor(0,1);
 lcd.print("Hello world ");
  delay(2000);


 if(!RTC.isrunning())
 {
 RTC.adjust(DateTime(__DATE__,__TIME__));
 }
}


void loop()
{
 int temp=0,val=1,temp4;
 DateTime now = RTC.now();
 if(digitalRead(set_mad) == 0)    //set Alarm time
 {
  lcd.setCursor(0,0);
  lcd.print(" Set Alarm ");
  delay(2000);
  defualt();
```

```
 time();
 delay(1000);
 lcd.clear();
 lcd.setCursor(0,0);
 lcd.print(" Alarm time ");
 lcd.setCursor(0,1);
 lcd.print(" has been set  ");
 delay(2000);
}
lcd.clear();
lcd.setCursor(0,0);
lcd.print("Time:");
lcd.setCursor(6,0);
lcd.print(HOUR=now.hour(),DEC);
lcd.print(":");
lcd.print(MINUT=now.minute(),DEC);
lcd.print(":");
lcd.print(SECOND=now.second(),DEC);
lcd.setCursor(0,1);
lcd.print("Date: ");
lcd.print(now.day(),DEC);
lcd.print("/");
lcd.print(now.month(),DEC);
lcd.print("/");
lcd.print(now.year(),DEC);
match();
delay(200);
}
void defualt()
{
```

```
 lcd.setCursor(0,1);
 lcd.print(HOUR);
 lcd.print(":");
 lcd.print(MINUT);
 lcd.print(":");
 lcd.print(SECOND);
}
/*Function to set alarm time and feed time into Inter-
nal eeprom*/
void time()
{
 int temp=1,minuts=0,hours=0,seconds=0;
  while(temp==1)
  {
  if(digitalRead(INC)==0)
  {
   HOUR++;
   if(HOUR==24)
   {
    HOUR=0;
   }
   while(digitalRead(INC)==0);
  }
  lcd.clear();
   lcd.setCursor(0,0);
  lcd.print("Set Alarm Time ");
 //lcd.print(x);
  lcd.setCursor(0,1);
  lcd.print(HOUR);
  lcd.print(":");
```

```
lcd.print(MINUT);
lcd.print(":");
lcd.print(SECOND);
delay(100);
if(digitalRead(next)==0)
{
 hours1=HOUR;
 EEPROM.write(add++,hours1);
 temp=2;
 while(digitalRead(next)==0);
}
}


  while(temp==2)
 {
 if(digitalRead(INC)==0)
 {
 MINUT++;
 if(MINUT==60)
 {MINUT=0;}
 while(digitalRead(INC)==0);
 }
 //lcd.clear();
 lcd.setCursor(0,1);
 lcd.print(HOUR);
 lcd.print(":");
 lcd.print(MINUT);
 lcd.print(":");
 lcd.print(SECOND);
 delay(100);
```

```
   if(digitalRead(next)==0)
   {
    minut=MINUT;
    EEPROM.write(add++, minut);
    temp=0;
    while(digitalRead(next)==0);
   }
  }
  delay(1000);
}
/* Function to chack medication time */
void match()
{
 int tem[17];
 for(int i=11;i<17;i++)
 {
  tem[i]=EEPROM.read(i);
 }
 if(HOUR == tem[11] && MINUT == tem[12])
 {
  beep();
  beep();
  beep();
  beep();
  lcd.clear();
  lcd.print("Wake Up........");
  lcd.setCursor(0,1);
  lcd.print("Wake Up.......");
  beep();
  beep();
```

```
 beep();
 beep();
 }
}
/* function to buzzer indication */
void beep()
{
 digitalWrite(buzzer,HIGH);
 delay(500);
 digitalWrite(buzzer, LOW);
 delay(500);
}
```

◆ ◆ ◆

4. RECURRENCE COUNTER UTILIZING ARDUINO

Pretty much every electronic specialist more likely than not confronted a situation where the person must gauge the recurrence of sign produced by a clock or a counter or a clock. We can utilize oscilloscope to carry out the responsibility, however not we all can bear the cost of an oscilloscope. We can purchase hardware for estimating the recurrence however all of these contraptions are expensive and are not for everybody.

We are gonna to plan a basic yet proficient Frequency Counter utilizing Arduino Uno as well as Schmitt trigger door.

This Frequency Counter is savvy and can be effectively made, we are going to utilize ARDUINO UNO for the estimating the recurrence of sign, UNO is the core of task here.

To test the Frequency Meter, we are going to make a spurious sign generator. This fake sign generator will be made by utilizing a 555 clock chip. The clock circuit produces a square wave which will be given to UNO to testing.

With everything set up we will have a Frequency meter and a square wave generator.

Required Components:

- 555 clock IC and 74LS14 Schmitt trigger entryway or NOT door.
- 100nF capacitor (2 pieces), 1000μF capacitor
- 1K ? resistor(2 pieces), 100? resistor
- 47K? pot,
- 16*2 LCD,
- Breadboard just as certain connectors.

Circuit Explanation:

The circuit outline of the Recurrence Meter utilizing Arduino is appeared in beneath figure. Circuit is basic, a LCD is interfaced with Arduino to show the deliberate recurrence of sign. 'Wave Input' is going to Signal Generator Circuit, from which we are sustaining sign to Arduino. A Schmitt trigger door (IC 74LS14) is utilized to guarantee that lone rectangular wave is sustained to Arduino. For separating the commotion we have included couple of capacitors crosswise over power. This Frequency Meter can gauge frequencies up to 1 MHz.

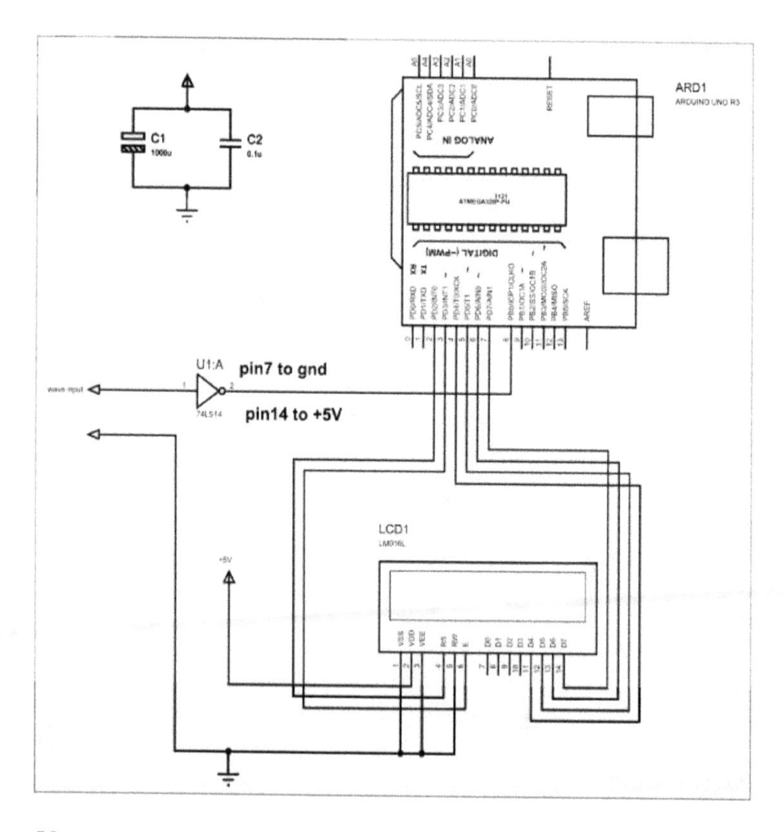

Signal generator circuit and Schmitt trigger have been clarified underneath.

Signal Generator using 555 Timer IC:

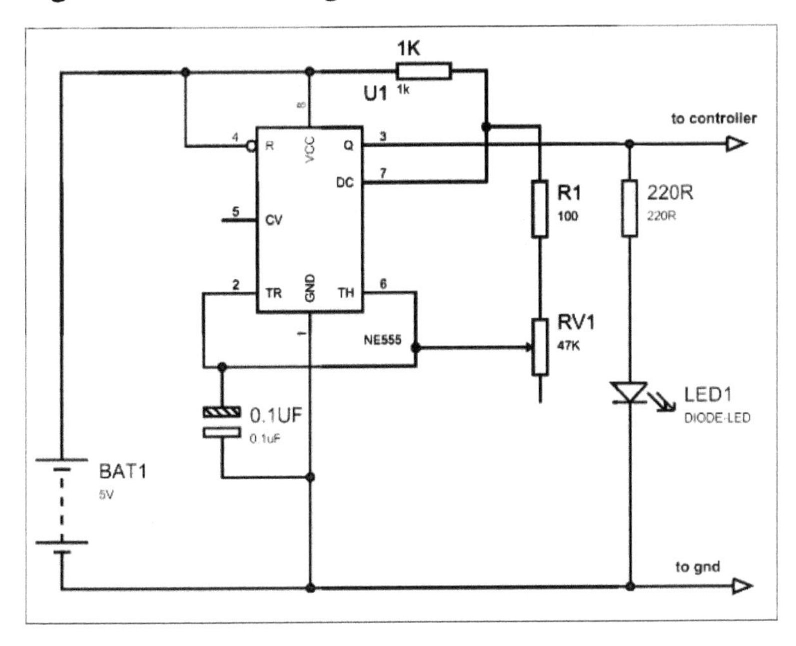

As a matter of first importance we will discuss 555 IC based square wave generator, or should I say 555 Astable Multivibrator. This circuit is essential on the grounds that, with the Frequency Meter set up we should have a sign whose recurrence is known to us. Without that sign we will always be unable to tell the working of Frequency Meter. In the event that we have a square have of known recurrence we can utilize that sign to test the UNO meter and we can change it for modifications for precision, if there should arise an occurrence of any deviations. The image of Signal Generator utilizing 555 Timer IC is given underneath:

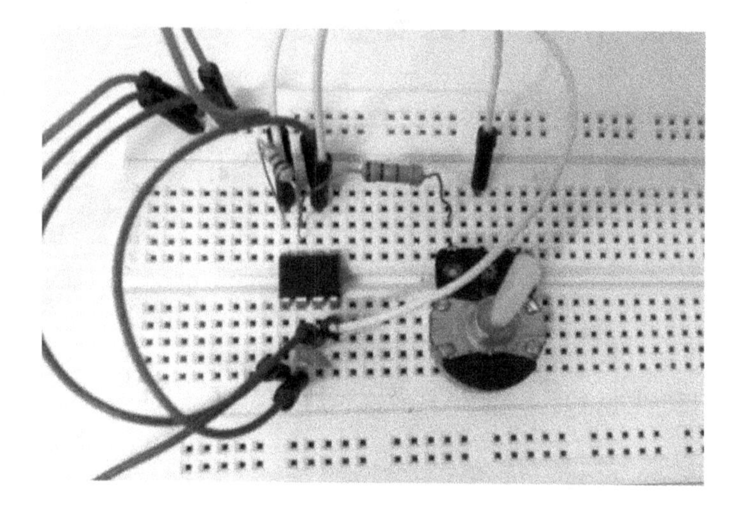

Common circuit of 555 in Astable mode is given beneath, from which we have inferred the above given Signal Generator Circuit.

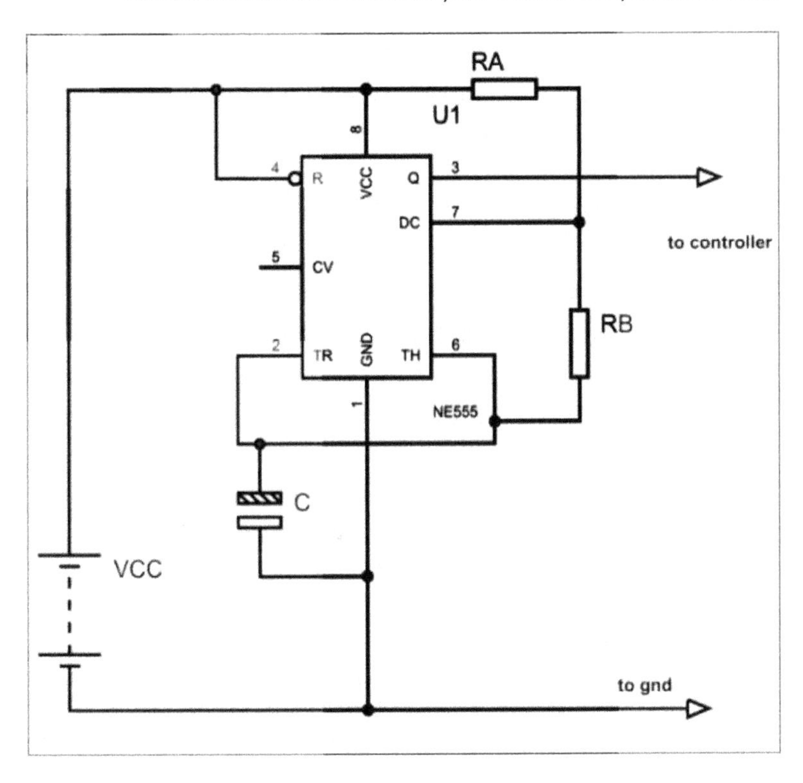

The yield signal recurrence relies upon RA, RB resistors and capacitor C. The condition is given as,

Recurrence (F) = 1/(Time period) = 1.44/((RA+RB*2)*C).

Here RA and RB are obstruction esteems and C is capacitance esteem. By putting the obstruction and capacitance esteems in above condition we get the recurrence of yield square wave.

One can see that RB of above graph is supplanted by a pot in the Signal Generator Circuit; this is done with the goal that we can get variable recurrence square wave at the yield for better testing. For effortlessness, one can supplant the pot with a basic resistor.

Schmitt Trigger Gate:

We realize that all the testing sign are not square or rectangular waves. We have triangular waves, tooth waves, sine waves, etc. With the UNO having the option to distinguish just the square or rectangular waves, we need a gadget which could change any sign to rectangular waves, in this way we use Schmitt Trigger Gate. Schmitt trigger entryway is a computerized rationale door, intended for number juggling and legitimate tasks.

This door gives OUTPUT dependent on INPUT voltage level. A Schmitt Trigger has a THERSHOLD voltage level, when the INPUT sign applied to the door has a voltage level higher than the EDGE of the rationale entryway, OUTPUT goes HIGH. In the event that the INPUT voltage sign level is lower than THRESHOLD, the OUTPUT of entryway will be LOW. We don't normally get Schmitt trigger independently, we generally have a NOT door following the Schmitt trigger. Schmitt Trigger working is clarified here: Schmitt Trigger Gate

We are going to utilize 74LS14 chip, this chip has 6 Schmitt Trigger entryways in it. These SIX entryways are associated inside as appeared in beneath figure.

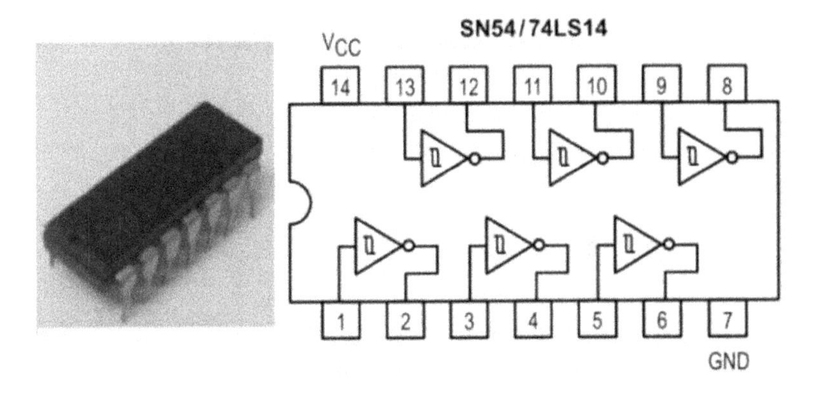

The Truth Table of Inverted Schmitt Trigger entryway is appear in beneath figure, with this we need to program the UNO for upsetting the positive and negative timespans at its terminals.

$Y = \bar{A}$

Input	Output
A	Y
L	H
H	L

H = High Logic Level

L = Low Logic Level

Presently we will nourish any sort of sign to ST door, we will have rectangular flood of transformed timeframes at the yield, we will bolster this sign to UNO.

Arduino measures the Frequency:

The Uno has an uncommon capacity pulseIn, which empowers us to decide the positive state term or negative state span of a specific rectangular wave:

```
Htime = pulseIn(8,HIGH);

Ltime = pulseIn(8, LOW);
```

The given capacity estimates the ideal opportunity for which High or Low level is available at PIN8 of Uno. So in a solitary cycle of wave, we will have the term for the positive and negative levels in Micro seconds. The pulseIn capacity estimates the time in smaller scale seconds. In a given sign, we have high time = 10mS and low time = 30ms (with recurrence 25 HZ). So 30000 will be put away in Ltime number and 10000 in Htime. At the point when we include them together we will have the Length Duration, as well as by reversing it we will have the Frequency.

Code

```
#include <LiquidCrystal.h>
LiquidCrystal lcd(2, 3, 4, 5, 6, 7);
int Htime;        //integer for storing high time
int Ltime;        //integer for storing low time
float Ttime;         // integer for storing total time of a
cycle
float frequency;     //storing frequency
void setup()
{
```

```
  pinMode(8,INPUT);
  lcd.begin(16, 2);
}
void loop()
{
  lcd.clear();
  lcd.setCursor(0,0);
  lcd.print("Frequency of signal");
  Htime=pulseIn(8,HIGH);   //read high time
  Ltime=pulseIn(8,LOW);    //read low time


    Ttime = Htime+Ltime;
    frequency=1000000/Ttime;     //getting frequency
with Ttime is in Micro seconds
  lcd.setCursor(0,1);
  lcd.print(frequency);
  lcd.print(" Hz");
  delay(500);
}
```

◆ ◆ ◆

5. INTERFACING ARDUINO WITH RASPBERRY PI UTILIZING SERIAL COMMUNICATION

Raspberry Pi and Arduino are the two most well known open source sheets in Electronics Community. They are mainstream among Electronics Engineers as well as among school understudies and specialists, due to their Easiness and Simplicity. Indeed, even a few people just began enjoying Electronics due to Raspberry Pi and Arduino. These sheets have extraordinary forces, and one can fabricate extremely convoluted and Hi-fi venture in hardly any basic advances and small programming.

We have made number of Arduino Projects and Tutorials, from straightforward ones to convoluted ones. We have additionally made Series of Raspberry Pi Tutorials, from where anybody can begin gaining 'without any preparation'. This is a little commitment towards Electronics Community from our side and this entryway has substantiated itself as Great Learning Resource for Electronics. So today we are bringing these two extraordinary sheets together by Interfacing Arduino with Raspberry Pi.

In this instructional exercise, we will build up a Serial Communication between Raspberry Pi as well as Arduino Uno. PI has just 26 GPIO sticks and zero ADC channels, so when we do undertakings like 3D printer, PI can't do every one of the cooperations alone. So we need more yield pins and extra capacities, for adding more capacities to PI, we build up a correspondence among PI and UNO. With that we can utilize all the capacity of UNO as they were PI capacities.

Arduino is a major stage for venture improvement, having numerous sheets like Arduino Uno, Arduino Pro smaller than usual, Arduino Due and so forth. They are ATMEGA controller based sheets intended for Electronic Engineers and Hobbyists. Despite the fact that there are numerous sheets on Arduino stage, yet Arduino Uno got numerous thanks, for its simplicity of doing ventures. Arduino based program improvement condition is a simple method to compose the program when contrasted with others.

Components Required:

Here we are utilizing Raspberry Pi 2 Model B with Raspbian Jessie OS as well as Arduino Uno. All the fundamental Hardware and Software prerequisites, in regards to Raspberry Pi, are recently talked about, you can find it in the Raspberry Pi Introduction, other than that we need:

- Associating pins
- 220? or 1K?resistor (2 pieces)
- Driven
- Catch

Circuit Explanation:

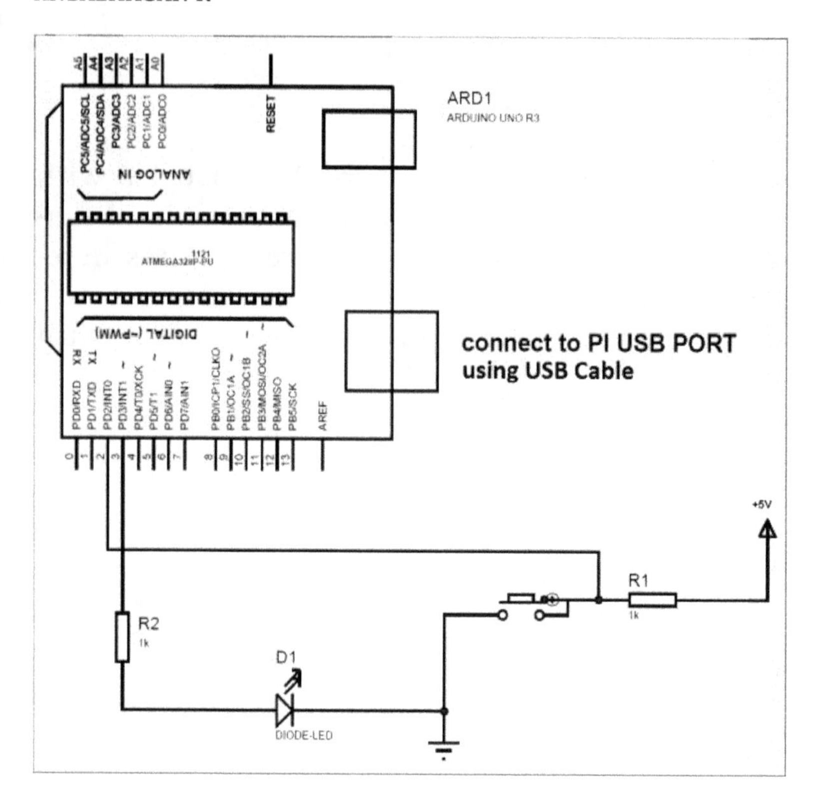

As appeared in the Circuit Diagram above, we will associate UNO to the PI USB port utilizing USB link. There are four USB ports for PI; you can interface it to any of them. A catch is associated with introduce the sequential correspondence and LED (squint) to demonstrate that information is being sent.

Working and Programming Explanation:

Arduino Uno Part:

First we should program the UNO,

Associate the UNO to the PC first and afterward compose the program (Check Code segment beneath) in the Arduino IDE programming and transfer the program to the UNO. At that point detach the UNO from PC. Join the UNO to the PI in the wake of programming and interface a LED and catch to the UNO, as appeared in circuit chart.

Presently the program here introduces the Serial Communication of UNO. At the point when we press the catch connected to the UNO, the UNO sends barely any characters to the PI sequentially through USB port. The LED connected to the PI squints to show the characters being sent.

Raspberry Pi Part:

After that we need to compose a program for PI (Check Code area beneath), to get this information being sent by UNO. For that we have to comprehend a couple of directions expressed underneath.

We are gonna to import sequential record from library, this capacity empowers us to send or get information sequentially or by USB port.

```
import serial
```

Presently, we have to express the gadget port and the bit rate for the PI to get the information from UNO with no mistakes. The underneath order expresses that, we are empowering the sequential correspondence of 9600 bits for each second on ACM0 port.

```
ser = serial.Serial('/dev/ttyACM0', 9600)
```

To discover the port which the UNO being joined to, go to the terminal of PI and enter

```
ls /dev/tty*
```

You will have the rundown of every single joined gadget on PI. Presently associate the Arduino Uno to Raspberry Pi with USB link and enter the direction once more. You can without much of a stretch distinguish the UNO appended port from the showed rundown.

Underneath direction is utilized as everlastingly circle, with this order the announcements inside this circle will be executed ceaselessly.

```
While 1:
```

Subsequent to accepting the information sequentially

we will show the characters on the screen of PI.

```
print (ser.readline())
```

So after the catch, joined to the UNO, is squeezed we will see characters being imprinted on the PI screen. Thus we have built up a Basic Communication Handshake between Raspberry Pi as well as Arduino.

Code

```
// Code for Arduino Uno
void setup()
{
pinMode(2,INPUT);                 // PIN2 is set for input
pinMode(3,OUTPUT);                // PIN3 is set for output
Serial.begin(9600);               // serial data rate is set for
9600bps(bits per second)
}
void loop()                       // execute the loop forever
{
 if(digitalRead(2)==LOW)          // if button attached to
the UNO is pressed
  {
  digitalWrite(3,HIGH);           // turn ON the LED at
PIN3
   Serial.println( "ButtonPressed" );   // send "Button-
Pressed" string of characters serially out
  delay(200);                     // wait for 200 milli second
  digitalWrite(3,LOW);            // turn OFF the LED
```

```
  }
}
```

 # Code for Raspberry PI

```
import time
import serial
ser = serial.Serial('/dev/ttyACM0', 9600)      # enable the
serial port
while 1:                          # execute the loop forever
ser.readline()                     # read the serial data
sent by the UNO
print (ser.readline())                        # print the serial
data sent by UNO
```

◆ ◆ ◆

6. MAKE YOUR OWN CUSTOM MADE ARDUINO BOARD WITH ATMEGA328 CHIP

Arduino is an open-source advancement stage for specialists and specialists to create hardware extends in a simple way. It comprises of both a physical programmable advancement board (in view of AVR arrangement of microcontrollers) as well as a bit of programming or IDE which runs on your PC and used to compose and transfer the code to the microcontroller board.

Arduino utilizes a boot loader. Boot loader is a bit of programming that enables the new programming to be scorched on it. So in this DIY, I will talk about "How to Consume a Boot Loader in a Crisp ATmega328 Chip as well as Build a Homemade Arduino on PCB". In Arduino UNO we use ATmega328 IC with the goal that I select this one to show this undertaking.

Components Required:

- Arduino UNO Board with IC just as link
- Breadboard
- Atmega328 IC
- 16 MHz precious stone oscillator
- Associating wires
- 10 K resistor

Steps for Building your own Arduino Board:

To consume a boot loader in new Atmega328 IC, we need an Arduino board (we can utilize any Arduino board to consume boot loader). And afterward we have to pursue beneath steps.

Stage 1. In initial step, orchestrate all the necessary things given in parts list above

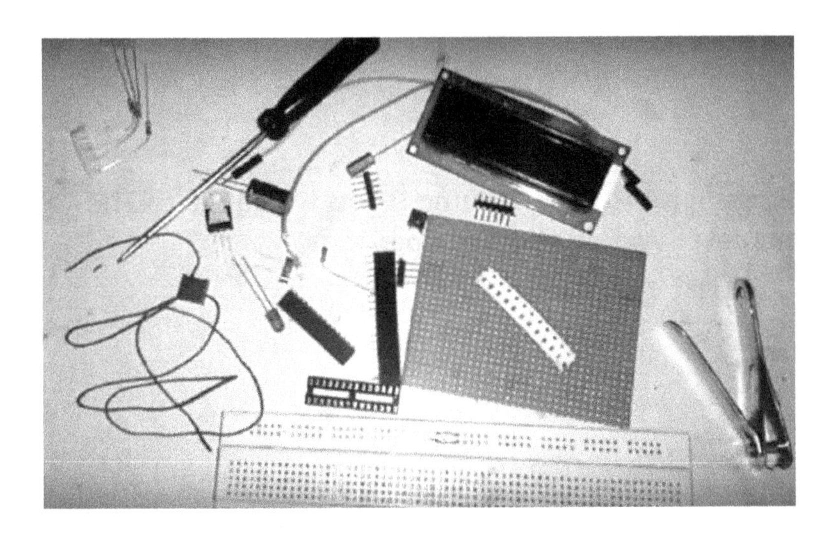

Stage 2: Now evacuate "Arduino Original IC" from Arduino board with the assistance of Screw Driver. What's more, embed "New Atmega328 IC" into the Arduino board.

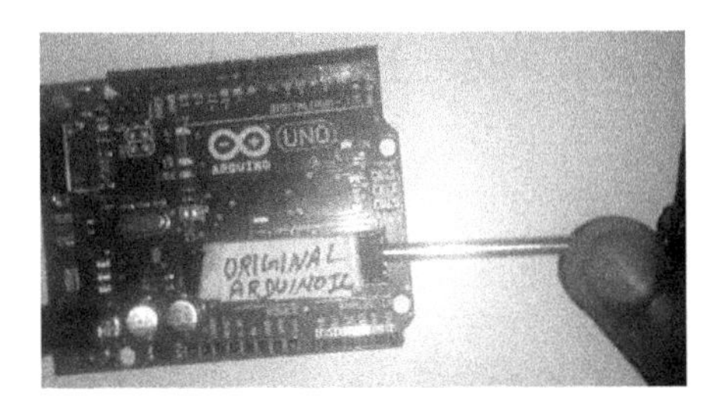

Stage 3: Now open Arduino IDE as well as go to File - > model - > ArduinoISP and open it.

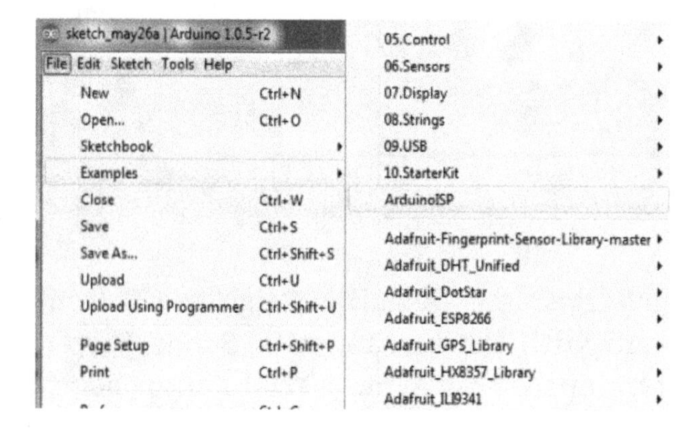

Subsequent to opening ArduinoISP, select Arduino UNO board from Apparatuses - > Board - > Arduino Uno.

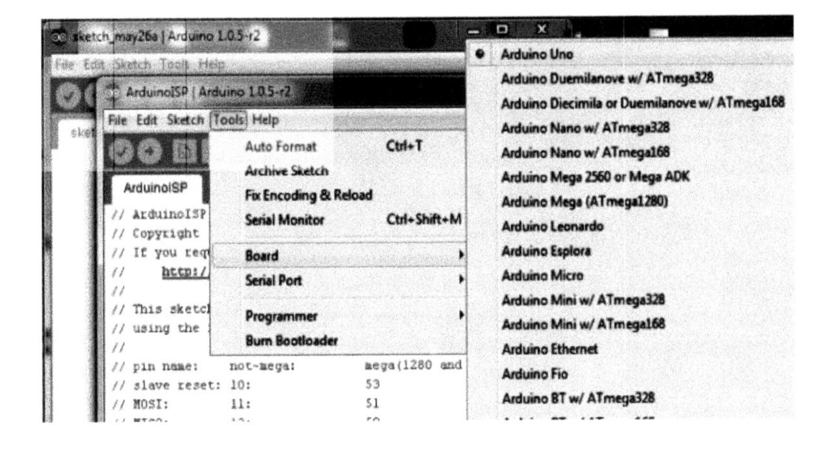

At that point select COM PORT from Tools - > Serial Port - > COM10

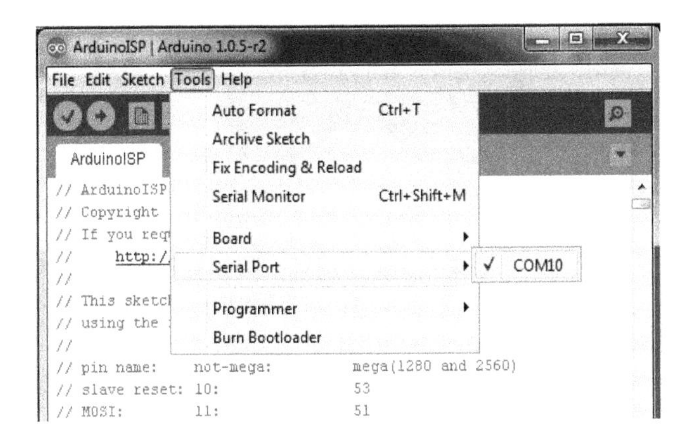

and afterward transfer ArduinoISP Sketch.

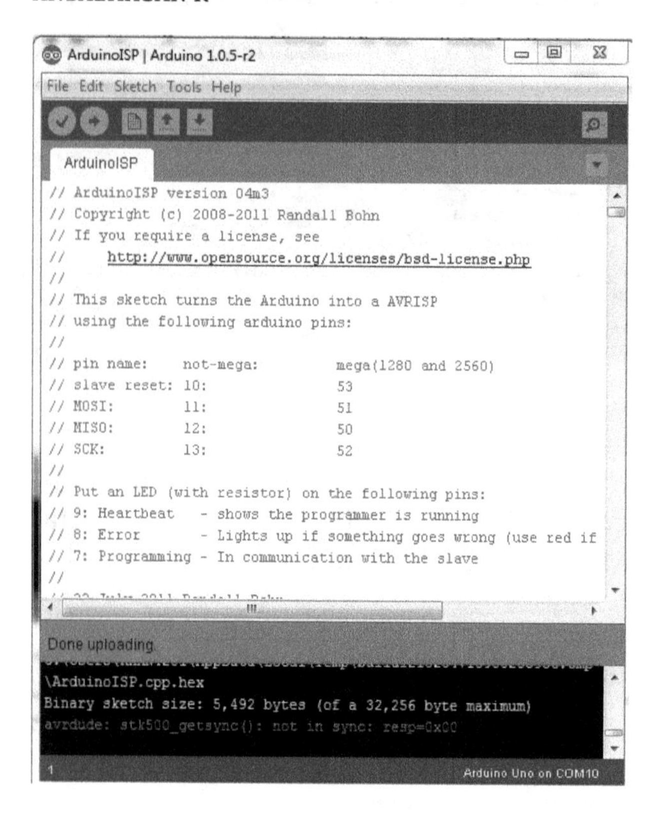

Stage 4: Now expel this New IC from the Arduino Board and addition the Arduino pre booted or Original Arduino IC into Arduino board and transfer the equivalent ArduinoISP sketch in it, similar to we have done in Step 3.

Stage 5: Build the beneath given Circuit on the bread board with New IC on Breadbaord as well as Original IC on Original Arduino Board.

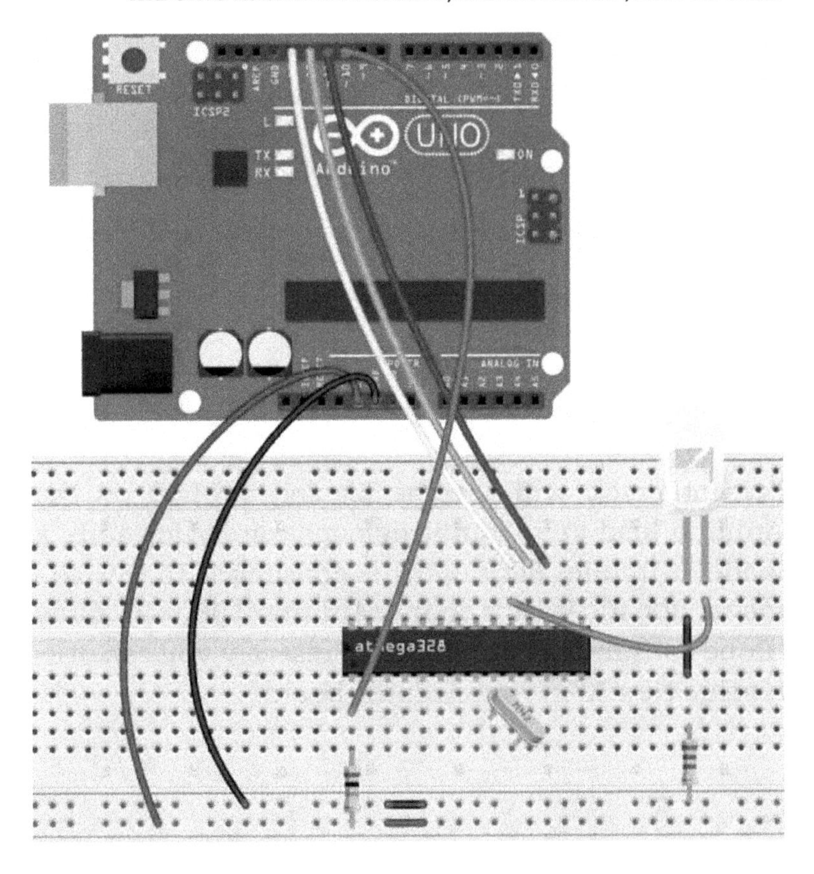

Stage 6: Now in Arduino IDE go to Tool as well as tap on the Burn Bootloader.

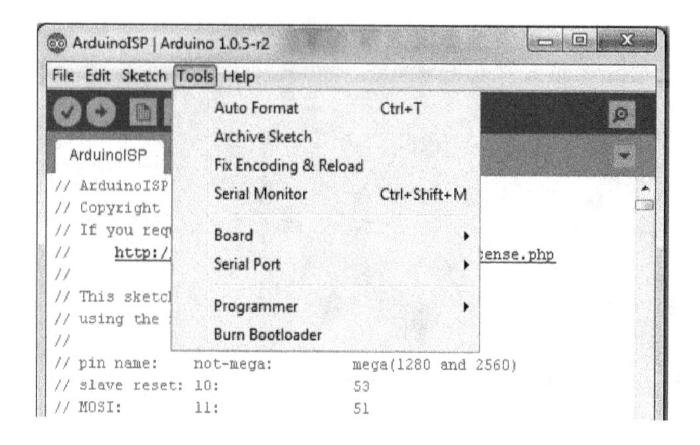

Presently you will see the Rx and Tx LED on the Arduino load up is squinting haphazardly for quite a while. It implies Bootloader is consuming in new AT-mega 328 IC. What's more, Arduino IDE will show "Done consuming bootloader". Presently you can utilize this 'New IC' in your Arduino board.

Stage 7: Now Build your own Homemade Arduino Board on Zero PCB by fastening the segments assem-

bled in Step 1, after the Circuit Diagram underneath.

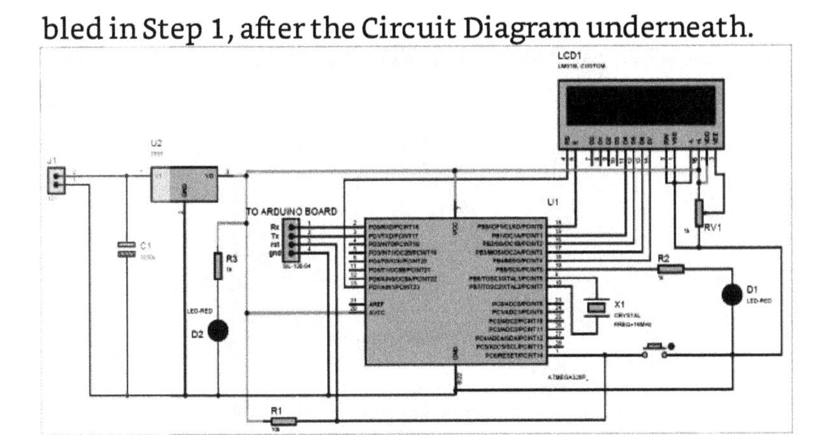

Supplement the 'New IC' in this board and you are finished.

You can likewise fabricate it appropriately on PCB with a legitimate PCB design and carving. Learn here to Make PCB at Home and convert Schematic into PCB format utilizing EasyEDA.

For LCD Interfacing, simply associate your home made Arduino Board with Original Arduino Board utilizing Rx, Tx, RST and GND pins of Original Arduino Board, as appeared in underneath Fritzing Circuit or above Circuit Diagram. Furthermore, transfer the Below Given (Code segment).

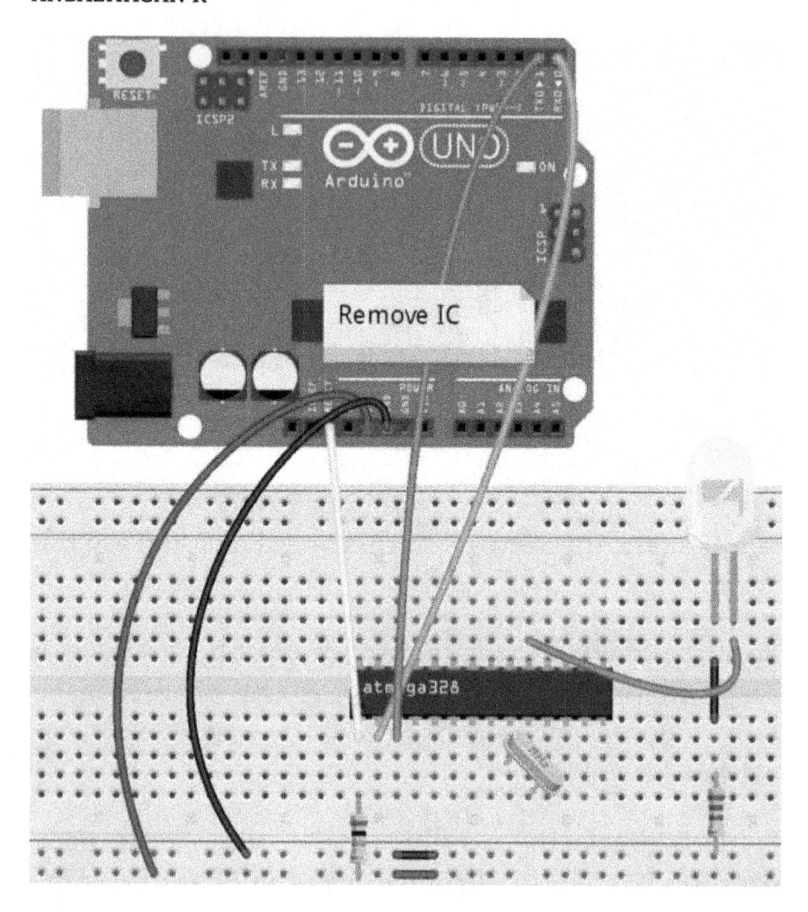

If you don't mind Remove 'Arduino Original IC' from the board, when you transfer code in new Arduino IC on the bread board or Zero PCB. You can control your Arduino Board with the 5v stick of Unique Arduino Board, as we have done in above Fritzing Circuit.

Code

```
#include <LiquidCrystal.h>
LiquidCrystal lcd(12, 11, 5, 4, 3, 2);
void setup()
{
 lcd.begin(16, 2);
 lcd.print("HomeMade Arduino");
 lcd.setCursor(0,1);
 lcd.print("Hello world");
}
void loop()
{


}
```

◆ ◆ ◆

7. INTERFACING TFT LCD WITH ARDUINO

Today, we are gonna to Interface 2.4 inch TFT LCD Shield with Arduino. By utilizing this shading TFT LCD shield we can show characters, strings, squares, pictures and so forth on the shading TFT LCD. Furthermore, we can utilize this TFT Shield in numerous applications like: Security Framework, Home Computerization, Games and so on.

Interfacing TFT LCD with Arduino is extremely simple. We just necessary to have an Arduino Board and a 2.4 inch TFT Shield in equipment part as well as Arduino IDE as well as TFT Library in programming part. Numerous libraries are accessible on the Internet, for TFT Shield to work, however extraordinary TFT LCDs have diverse inbuilt drivers. So first we have to recognize the driver of TFT and afterward introduce an appropriate library for that. Here we are utilizing 2.4 inch TFT Shield having ili9341 driver. Connection, for installing the library for given TFT, is yielded 'Steps' underneath. Check this for basic LCD interfacing with Arduino.

Hardware and Software Requirements:

Hardware:

- Arduino Uno
- TFT shield

- USB cable

Software:

- TFT library for Arduino (spfd5408)
- Arduino ide

Circuit Diagram:

Client just needs to embed TFT Shield over the Arduino. Since TFT Shield is perfect with Arduino UNO and Arduino mega.

Steps for Installing TFT library in Arduino **IDE:**

Stage 1: Download the TFT library for Arduino, from the underneath given connection and make it zip (if not as of now zipped).

https://github.com/JoaoLopesF/SPFD5408

Stage 2: After this, reorder it in Arduino library organizer in Program Files.

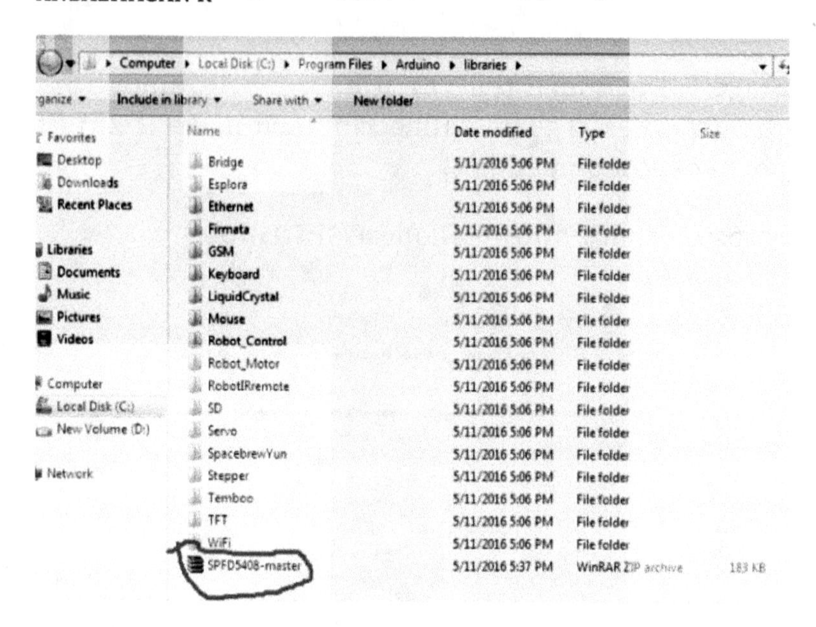

Stage 3: Now open Arduino IDE and select Sketch - > Include Library - > Add .ZIP Library.

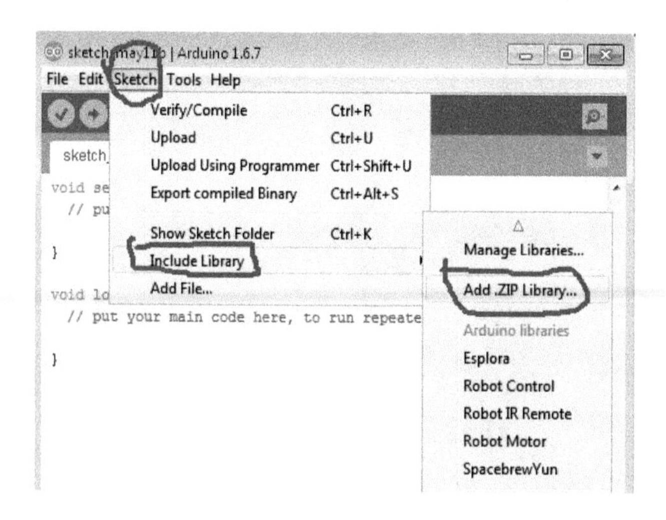

At that point go to the Arduino Library in Program Files, where you have stuck the compressed downloaded library in Step 2 and select and open compressed SPFD5408-Master library.

My Computer - > C: Drive - > Program Files - > Arduino - > libraries

Presently subsequent to opening the SPFD5408 Ace Library, you can notice that your library document has been introduced in Arduino IDE.

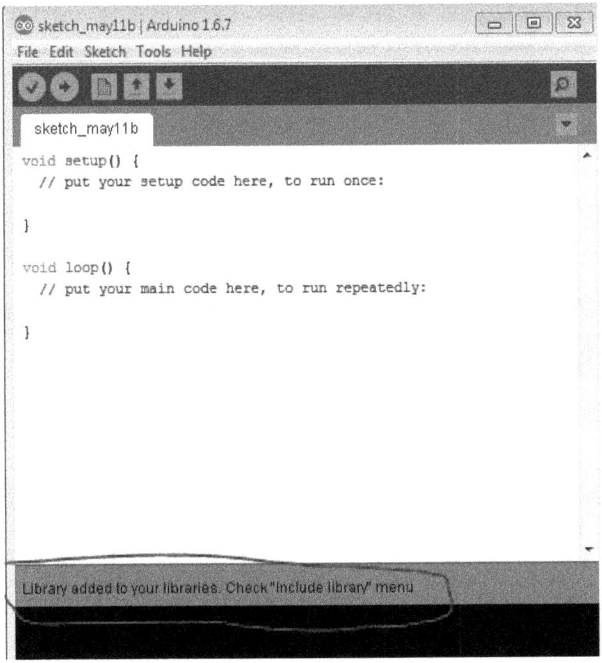

sketch_may11b | Arduino 1.6.7

File Edit Sketch Tools Help

sketch_may11b

```
void setup() {
  // put your setup code here, to run once:

}

void loop() {
  // put your main code here, to run repeatedly:

}
```

Library added to your libraries. Check "Include library" menu

Stage 4: Now in Arduino IDE go to, File - > Example - > SPFD5408-ace - > spfd5408_graphictest

Open it, incorporate it and afterward transfer it in Arduino.

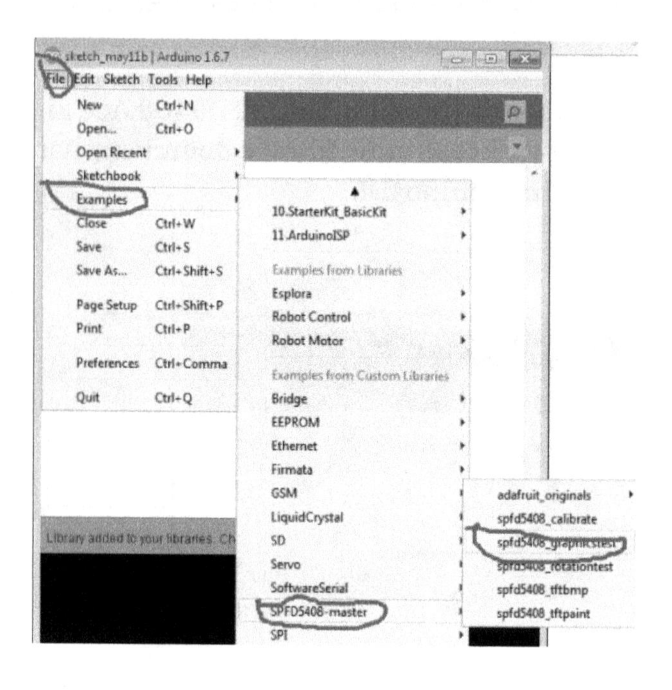

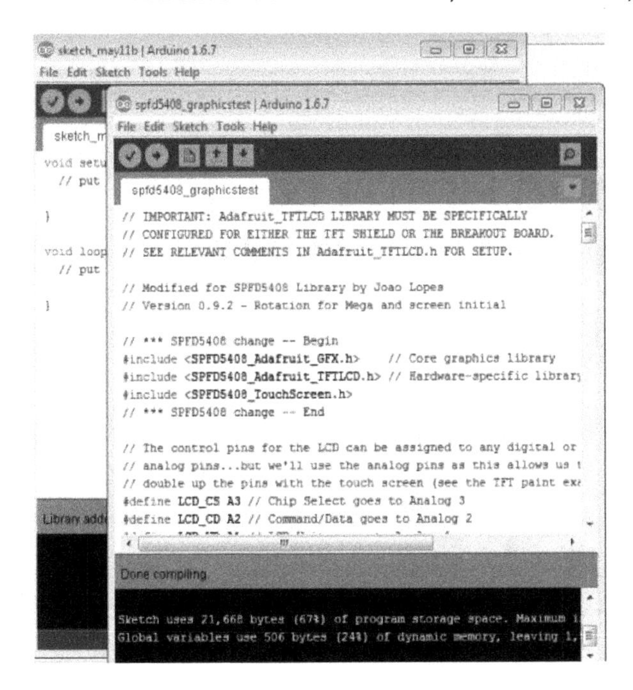

Presently you will get your outcomes on TFT. Clients can change this code as indicated by their necessities, similar to we have embedded some 'Content' as per us.

Note: Same advances can be pursued to introduce any Library in Arduino IDE.

8. SERVO MOTOR CONTROL WITH ARDUINO DUE

As examined before, Arduino Due is an ARM controller based board intended for electronic architects and specialists. This DUE board can be utilized for making CNC machines, 3D printers, automated arms and so on. Every one of these undertakings have a typical component that is Position Control. Any of these ventures needs precision towards their position. Exact positions in these machines can be accomplished by Servo Motors. In this session we are going to control the situation of a Servo Motor with Arduino Due. We have just secured the Servo Motor Interfacing with Arduino Uno as well as Servo Motor Interfacing with 8051 Microcontroller.

Servo Motors:

Servo Engines are known for their precise shaft development or then again position. These are not proposed for rapid applications. These are proposed for low velocity, medium torque as well as precise position application. These engines are utilized in mechanical arm machines, flight controls and control frameworks. Servo engines are additionally utilized in some of printers and fax machines.

Servo engines are accessible at various shapes and sizes. We will utilize SG90 Servo Motor in this instructional exercise. SG90 is a 180 degree servo engine. So with this servo we can situate the pivot from 0 to 180 degrees.

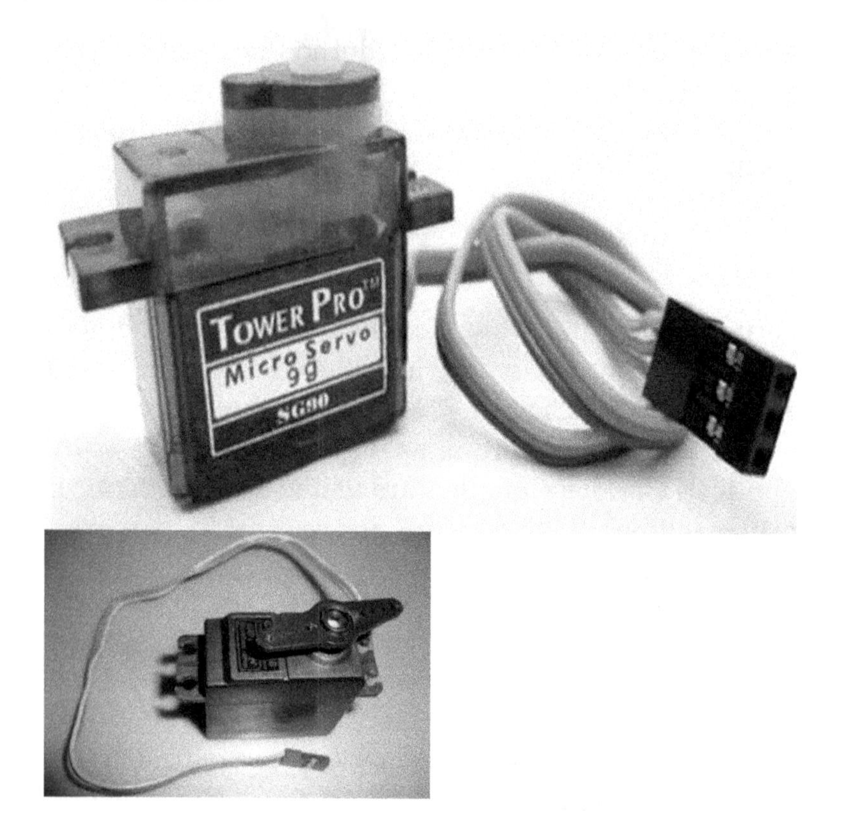

A Servo Motor for the most part has three wires, one is for positive voltage, another is for ground and last one is for position setting. The RED wire is associated with control, Brown wire is associated with ground and YELLOW wire (or WHITE) is associated with signal.

A Servo Motor is a mix of DC engine, position control framework and riggings. In servo, we have a control framework which takes the PWM signal from sign stick. It translates the sign and gets the obligation proportion from it. After that it thinks about the proportion to the predefined positions esteems. In case there is a distinction in the qualities, it changes the situation of the servo likewise. So the pivot position of the servo engine depends on the obligation proportion of the PWM sign to the SIGNAL stick.

The recurrence of PWM (Pulse Width Modulated) sign can change dependent on sort of servo engine. The significant thing here is the DUTY RATIO of the PWM signal. Check this for PWM with Arduino Due. Anyway for this situation, we need not stress over with the Duty Ratio determination. In Arduino we have an exceptional capacity; after calling it we can alter the situation of servo, just by expressing the edge. We will discuss that in the Working Section beneath.

Before Interfacing Servo Engine to Arduino Due, you can test your servo with the assistance of this Servo Engine Tester Circuit. Additionally check these tasks to Control Servo by Flex Sensor or by Force Sensor.

Components:

Equipment: Arduino Due, control supply (5v), Servo engine.

Programming: Arduino daily, download it from connect beneath (https://www.arduino.cc/en/Main/Software)

For, subtleties on How to download and introduce this

product, visit the main instructional exercise Getting Started with Arduino Due.

Circuit Diagram and Working Explanation:

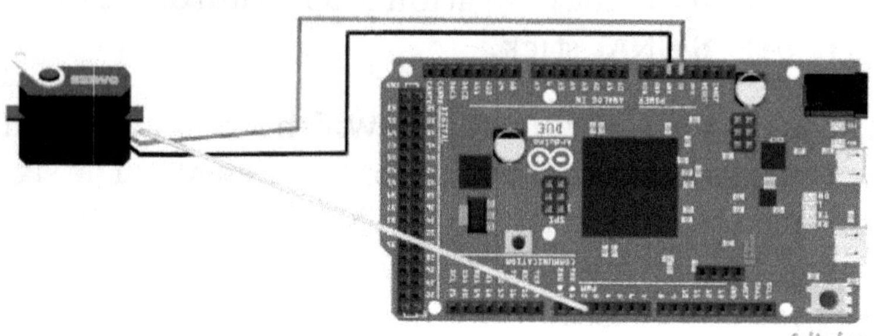

As said before in ARDUINO, we have predefined libraries, which will set the frequencies and obligation proportions as needs be, when the header document is called or included. In ARDUINO we basically need to express the situation of servo that required and the DUE produces proper PWM signal for the servo. The things which we have to accomplish for getting exact situation of servo are,

```
#include <Servo.h>
```

```
Servo myservo;
```

```
myservo.attach(servo_signal_pin_attached_to);
```

```
myservo.write(needed_position_ angle);
```

The header record "#include <Servo.h>" incorporates all the extraordinary capacities we need, after considering it we never again need to worry over the recurrence of PWM nor about the DUTY RATIO of sign. With this the client can enter required situation of servo straightforwardly with no fluff.

Furthermore a name is to be picked for the servo by utilizing "Servo myservo", here myservo is the name picked, so while composing for position we are going to utilize this name, this component proves to be useful when we have numerous servos to control, we can control upwards of 12 servos by this.

With the Arduino Due having 12 PWM channels, we have to disclose to DUE where the sign stick of servo is associated or where it needs to create the PWM signal. To do this we have "myservo.attach (2);", here we are telling the DUE that we have associated the sign stick of servo at PIN2.

All left is to set the position, we are gonna to set the situation of servo by utilizing "myservo.write(45);", by this order the servo hand moves 45 degrees. In the event that we change '45' to '175', the servo hub points to 175 degrees and remains there. From that point forward, at whatever point we have to change the situation of servo we simply need to call the direction "myservo.write(needed_position_angle);".

In the program, we are going to addition and decrement the points by utilizing circles. So the servo compasses from 0 to 180, at that point from 180 to 0, etc. The Servo Motor Control by Arduino Due is best clarified in bit by bit of C code down beneath.

Code

```
#include <Servo.h>
Servo myservo;          // providing a name
int angle = 0;                // variable to store the servo
position
void setup() {
 myservo.attach(2);     // attaches the servo on pin 2 to
the servo object
}
void loop() {
 for (angle = 0; angle <= 180; angle += 1) {   // goes from
0 degrees to 180 degrees, in steps of 1 degree
  myservo.write(angle);                    // tell servo to go
to position in variable 'angle'
  delay(15);                          // waits 15ms for the
servo to reach the position
 }
 for (angle = 180; angle >= 0; angle -= 1) {   // goes from
180 degrees to 0 degrees
```

```
 myservo.write(angle);
 delay(15);                    // waits 15ms for the
servo to reach the position
 }
}
```

◆ ◆ ◆

9. PWM WITH ARDUINO DUE

Arduino Due is an ARM controller based board intended for electronic architects and specialists. ARM design is extremely persuasive in present day gadgets, we use them wherever like our mobiles, iPods and PCs and so forth. In the event that somebody needs to structure mechanical frameworks it should on ARM controllers. ARM controllers are significant in view of their readiness.

We have just secured the essentials of Arduino Due in

Getting Started with Arduino Due. Presently in this instructional exercise we will modify the brilliance of a LED, by utilizing PWM sign created by DUE. A DUE PWM signal gives a variable voltage over consistent power supply.

Pulse Width Modulation:

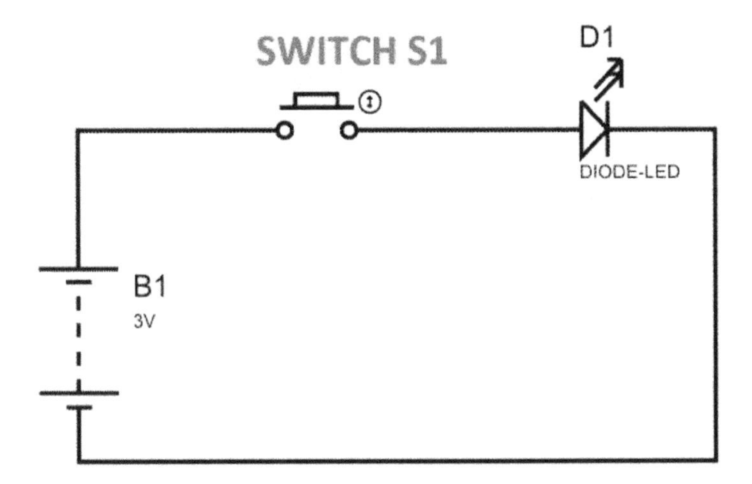

In above figure, if the switch is shut persistently over some undefined time frame, the LED will be 'ON' during this time ceaselessly. On the off chance that the switch is shut for half second and opened for next half second, at that point LED will be ON just in the main half second. Presently the extent for which the LED is ON over the all out time is known as the Duty Cycle, and can be determined as pursues:

Obligation Cycle =Turn ON schedule/(Turn ON time + Turn OFF time)

Obligation Cycle = (0.5/(0.5+0.5)) = half

So the normal yield voltage will be half of the battery voltage.

This is the situation for one second as well as we can notice the LED being OFF for half second as well as LED being ON the other half second. On the off chance that Frequency of ON and OFF occasions expanded from '1 every second' to '50 every second'. The human eye can't catch this recurrence of ON and OFF. For a typical eye the LED will be seen, as sparkling with half of the brilliance. So with further decrease of ON time the LED shows up a lot lighter.

We will program the DUE for getting a PWM and associate a LED to show its working.

There are 12 PWM Channels (Pin 2 to Pin 13) in the DUE and we can utilize any one or every one of them. For this situation we will adhere to one PWM signal at PIN2.

Components:

- Arduino Due
- 220? resistor
- LED
- Power supply (5v)
- 1K? resistor (two pieces),
- Buttons (two pieces).

Also, Arduino IDE - Arduino Nightly Software (https://www.arduino.cc/en/Main/Software).

Circuit Diagram and Working Explanation:

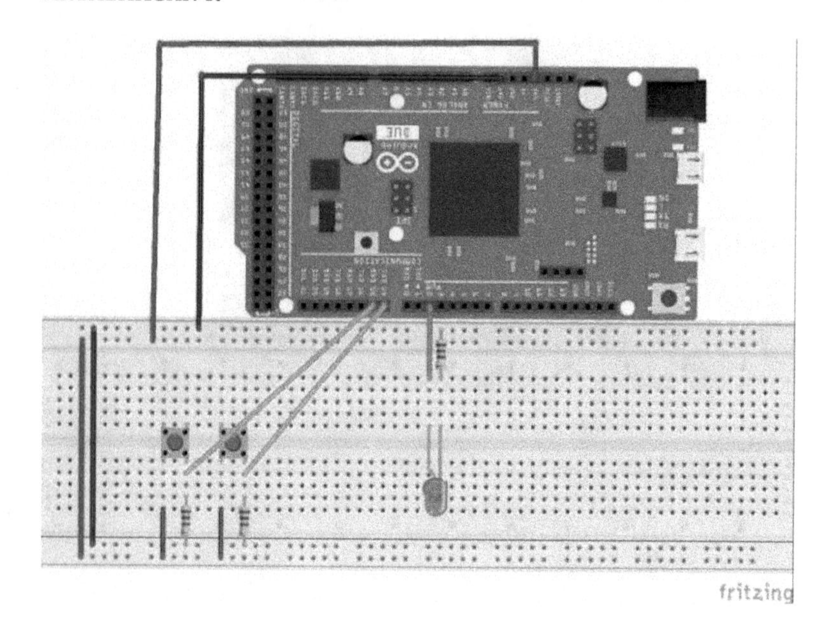

The circuit is associated on breadboard according to the Circuit Diagram. Anyway one must focus during interfacing the LED terminals. In spite of the fact that the catches can show ricocheting impact yet for this situation it doesn't cause extensive mistakes, so we need not stress this time.

Getting a PWM signal from DUE is simple; Arduino IDE gives valuable highlights which facilitates the software engineer's trouble. On the off chance that we go for exposed chip programming, we need setting up an AT-MEGA controller for PWM signal, which isn't simple; we need to characterize numerous registers and settings for a precise sign, anyway in Arduino we don't need to manage every one of those things. We have just secured Pulse width Modulation with ATmega32 ,with Arduino Uno and with 555 clock IC.

Naturally all the header documents and registers are predefined by Arduino IDE, we basically need to call them and that is it, we will have a PWM yield at suitable stick. We additionally need to call certain directions to get a PWM signal, these are examined underneath:

```
pinMode(2,OUTPUT)

analogWrite(pin, value)
```

First we have to pick the PWM yield channel or select a stick from 12 pins of DUE, after that we have to set that stick as yield. Since we are utilizing PIN2 as yield, we will set it as OUTPUT as appeared in first line.

Next we have to empower the PWM highlight of DUE by calling the capacity "analogWrite(pin, esteem)". In here 'stick' speak to the stick number where we need PWM yield. We are putting it as '2', so at PIN2 we are getting PWM yield. "Worth" is the turn ON esteem, it differs between 0 (constantly off) and 255 (consistently on). We can compose the suitable incentive in this space for required splendor of LED.

We connected two or three catch to the DUE board for shifting this worth. One catch is for increasing the brilliance worth and other is for diminishing the splendor esteem. When the Due is finished programming, we can alter splendor by squeezing these catches.

Code

```
volatile int brightness = 0;    //initializing a integer for incrementing and decrementing duty ratio.
void setup()
{
 // declare pin 2 to be an output:
 pinMode(2, OUTPUT);
 pinMode(14, INPUT);
 pinMode(15, INPUT);
}
// the loop routine runs over and over again forever:
void loop()
{
 if(digitalRead(14)==HIGH)
 {
  if(brightness<255)
  {
    brightness++;       //if pin14 is pressed and the duty ratio value is less than 255, increase the 'brightness'
  }
  delay(20);
 }
  if(digitalRead(15)==HIGH)
 {
  if(brightness>0)
  {
```

```
  brightness--;     //if pin15 is pressed and the duty ratio value is greater than 0, decrement the 'brightness'
  }
  delay(20);
 }
 analogWrite(2, brightness);
}
```

◆ ◆ ◆

10. WIDESPREAD IR REMOTE CONTROL UTILIZING ARDUINO AND ANDROID APP

I began this task so as to dispose of different remotes at my home and construct something single which could consolidate highlights of every one of them. I got this thought when I saw one of my companion's cell phone with worked in IR blaster, around then I chose not to purchase a comparative telephone rather make my own gadget which ought to be perfect with my current handset. We are gonna to Change over an Android Telephone into an IR Remote utilizing Arduino to control different gadgets at home.

Components Required:

- Arduino Uno
- TSOP-IR beneficiary (1838T)
- Bluetooth module (HC05)
- IR LED
- Android Device (Phone, Tablet, and so forth.)

Working Explanation:

By and large we utilize two remotes to work TV at home, one for TV and one for Set-Top Box so here in this Project I am focusing on these two remotes and making an Android Phone functioning as IR Blaster so TV can be controlled with the Phone, without contacting any of the Remotes.

Utilizing an Arduino Uno board simply made it simpler for me to manage the IR deciphering and encoding part. The hand crafted shield just includes to the accommodation part of this undertaking. The shield comprises of a TSOP IR recipient (1838T), an IR LED and a Bluetooth module (HC-05), see the picture underneath:

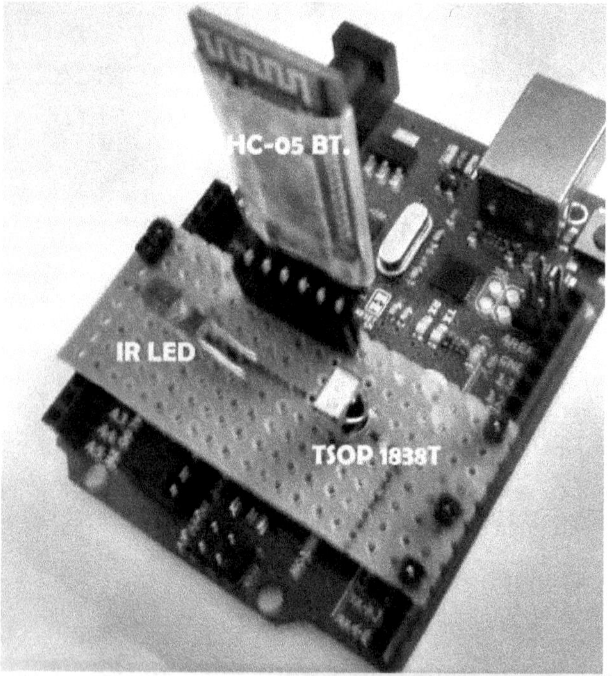

You can either assemble the custom shield or can legitimately interface the parts to the Arduino as appeared in the 'Circuit Diagram' in underneath area.

Before pushing ahead let us initially talk about 'how

the IR remotes work'. The greater part of the IR remotes work around 38 KHz frequencies (this is the motivation behind why I have picked 1838T). On further including into this subject one will perceive that there's no fixed portrayal for zeros and ones in these IR information transmission strategies. These codes utilize different encoding systems which we study in our designing schedule (since I am a gadgets building understudy). The hugeness of 38 KHz is that it is the recurrence at which sign wavers when sensibly high for example this is the bearer recurrence of the sign. Examine the image underneath; this is a case of NEC Protocol. This will make your idea all the more clear:

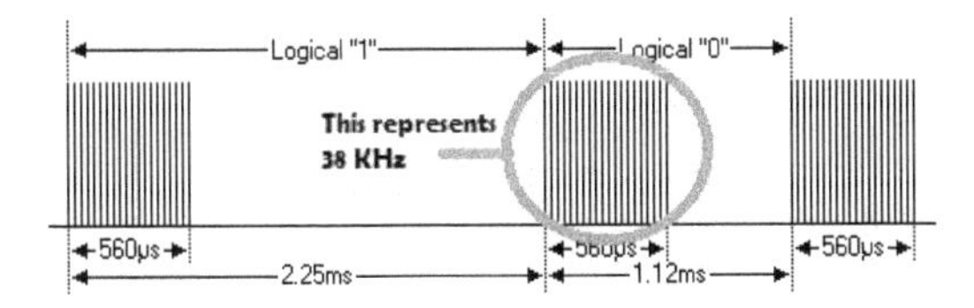

So here's the manner by which this IR Blaster works; an Android Phone with the specially crafted Android App sends the sign to Arduino circuit over Bluetooth, further the Arduino gets the sign through TSOP-IR recipient (1838T) and examinations it. At that point Arduino directions the IR LED to flicker in a specific example, comparing to the catch pushed on that Android Device App. This flickering example is caught by TV or Set-Top box's IR recipient and it adheres to the guidance appropriately like changing the channel or expanding the volume.

In any case, before that we have to decipher the current remotes. As referenced before, in this task I have utilized two remotes, one which speaks with the TV while another is for the Set-top box associated with TV.

Circuit Diagram:

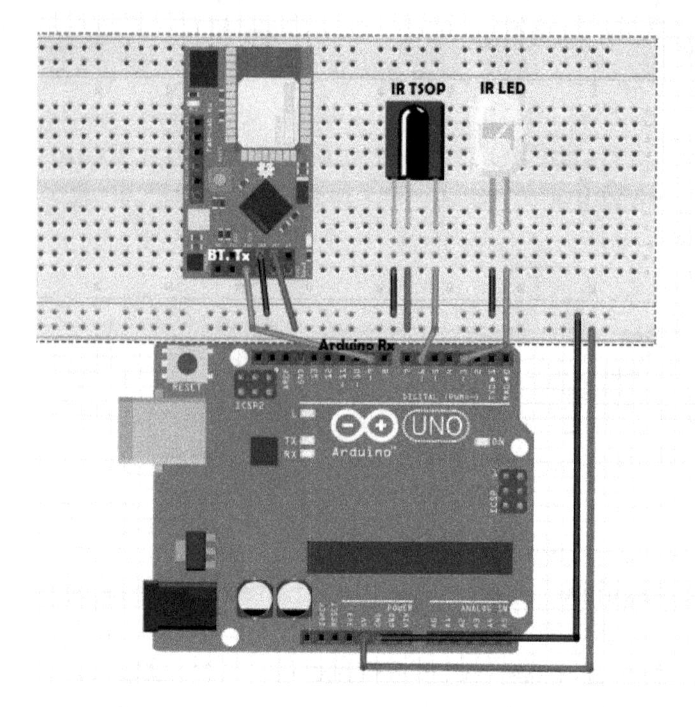

Decoding IR Remote Control Signals using Arduino:

The Arduino board here works in two stages, one is the point at which you are utilizing it to interpret IR codes from remote and another is the point at which you are utilizing it as the IR blaster gadget.

Give us a chance to discuss the principal stage. Presently to unravel the IR catch codes, I have utilized Ken Shirriff's IRremote header record. This header document has numerous predefined models/codes just to make it simpler for us to work with IR codes:

- You first need to download and introduce the IR remote library from here https://github.com/z3t0/Arduino-IRremote.
- Unfasten it, and spot it in your Arduino 'Libraries' envelope. At that point rename the removed organizer to IRremote.
- At that point consume the beneath gave code into Arduino, module the custom shield as appeared above and place a remote to be decoded before the TSOP IR collector. Open up the sequential screen comparing to this Arduino and press any ideal catch from the remote. You'll see some data showed over the terminal, this data includes the kind of code, its worth and the measure of bits engaged with it. Here's what it looks like:

```
#include <IRremote.h>

const int RECV_PIN = 6;
```

```
IRrecv irrecv(RECV_PIN);

decode_results results;

void setup()

{

 Serial.begin(9600);

 irrecv.enableIRIn(); // Start the receiver

 irrecv.blink13(true);

}

void loop() {

 if (irrecv.decode(&results)) {

  if (results.decode_type == NEC) {

   Serial.print("NEC ");

  } else if (results.decode_type == SONY) {

   Serial.print("SONY ");

  } else if (results.decode_type == RC5) {
```

```
Serial.print("RC5 ");
} else if(results.decode_type == RC6) {
Serial.print("RC6 ");
} else if(results.decode_type == UNKNOWN) {
Serial.print("UNKNOWN ");
}
Serial.print(" ");
Serial.print(results.value, HEX);
Serial.print(" ");
Serial.println(results.bits);
irrecv.resume(); // Receive the next value
}
}
```

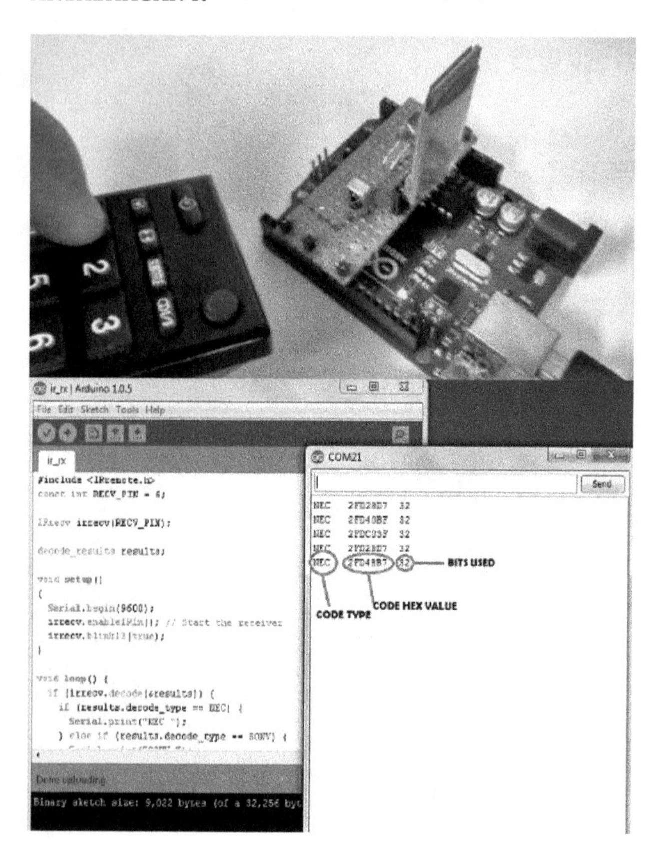

When you are finished with the ideal deciphering part, note down all the interpret values and other data with their relating catch name squeezed. This will fill in as a database for the following period of Arduino. The above program is taken from IRremote library's 'models' organizer, you can look at more guides to become familiar with utilizing the IR remote. With the goal that's the means by which we decoded the IR remote yield.

Presently consume the Code, given in Code area toward the end, onto this equivalent board. Congrats, you are finished with the principal half of this venture.

Building the Android App for IR Blaster:

Here comes the subsequent a large portion of, the Android App making. I basically incline toward utilizing MIT's APP designer 2 for making such sort of applications. In case you are a novice in Android coding, this will spare your time and give great outcomes. The principle parts utilized in creation of this application are very little, only not many catches and a Bluetooth customer bundle. While coding the application, give the comparing content to be sent for each catch pushed on the screen which would ask Arduino to squint IR LED in a similar way as it would have been finished by the individual remote; additionally, ensure that you give the right address of your Bluetooth HC-05 module. This is the means by which the last App will look in your Android Smart Phone:

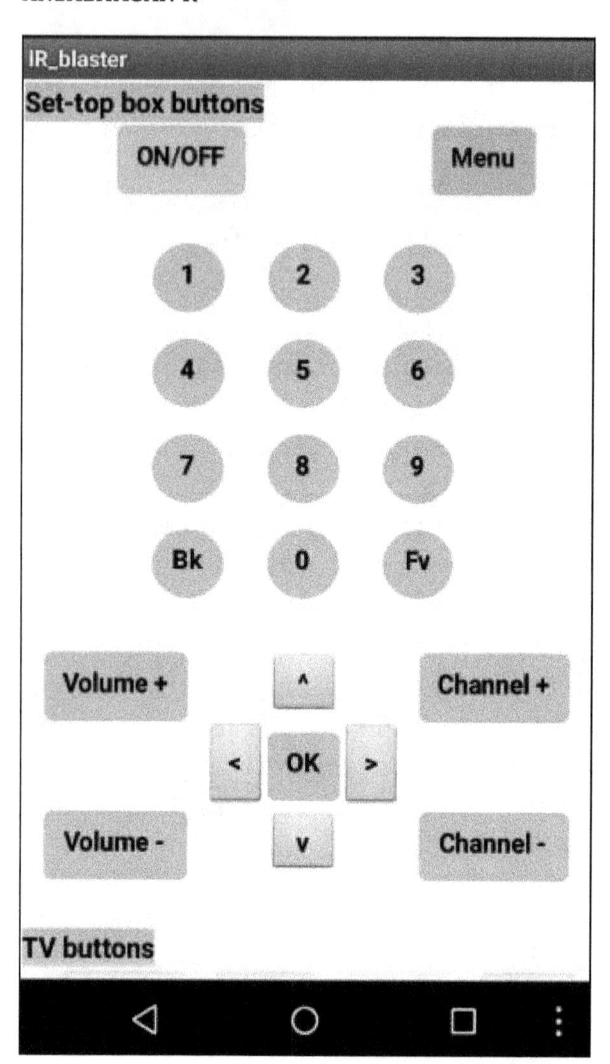
IR_blaster
Set-top box buttons
ON/OFF
Menu
1
2
3
4
5
6
7
8
9
Bk
0
Fv
Volume +
^
Channel +
<
OK
>
Volume -
v
Channel -
TV buttons

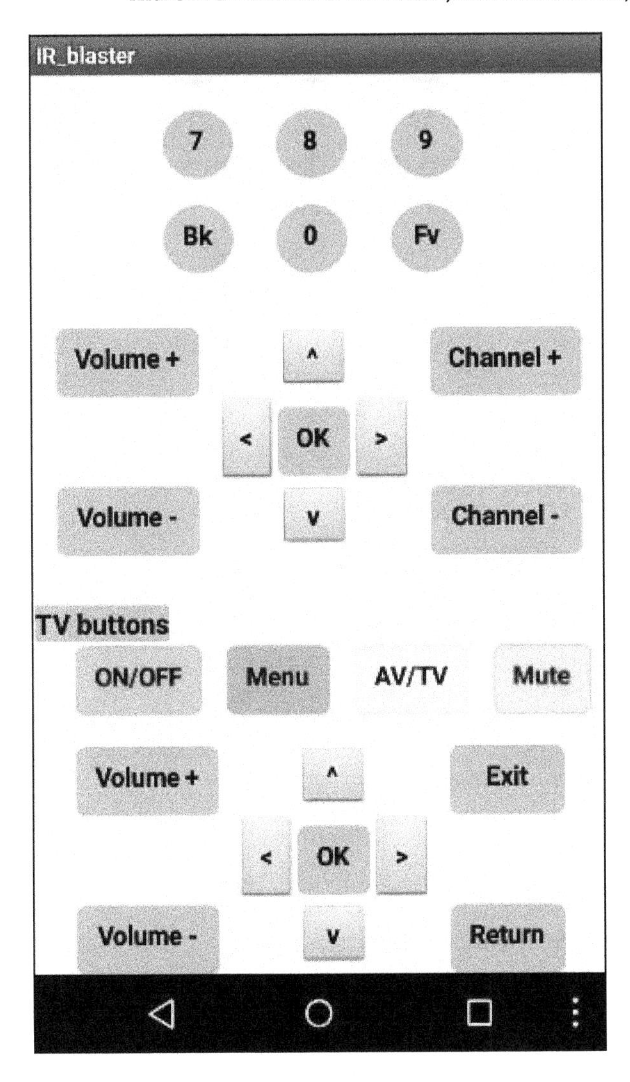

Here is the bit by bit procedure to manufacture the application:

Stage 1:

Sign on to this connection: ai2.appinventor.mit.edu, or attempt and search out MIT appinventor-2 on Google. Marking in to AI2 requires a Google account, so on the off chance that you don't have, make one.

Stage 2:

When you sign in with your Google account, you'll be diverted to AI2 working site page, which resembles this:

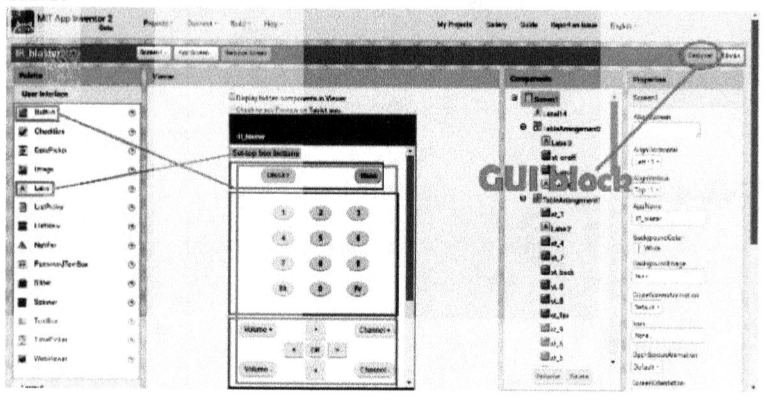

Start another task by clicking "Activities" tab at the top and select "Start New Project". At the point when you are finished with the naming part and each of the, a clear screen will be shown to you in which you can place catches and content as appeared previously. This is the GUI screen, in which you choose how the application might look want to a client.

To utilize a catch bundle, select "Catch" tab on the left half of the screen under "UI" area. Simply drag any bundle from left-side menu and drop it onto the working screen. Also to show any content related stuff, utilize "Name" bundle.

Stage 3:

In the wake of orchestrating every one of your catches and marks, presently it's an ideal opportunity to make a code for this application. In any case, before that we have to choose a Bluetooth bundle too for speaking with the Arduino.

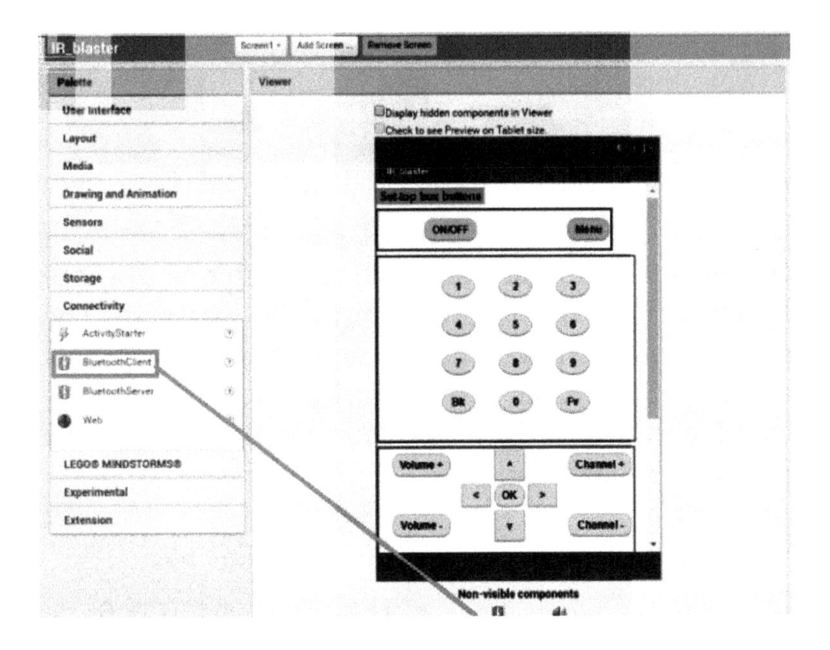

You'll see that this bundle isn't shown on the screen rather it goes under "Non-unmistakable Components". These are the parts that have no criticalness in GUI make-up.

Stage 4:

Next comes the coding area, wherein you'll characterize the capacity for parts that you have chosen and you need to work with.

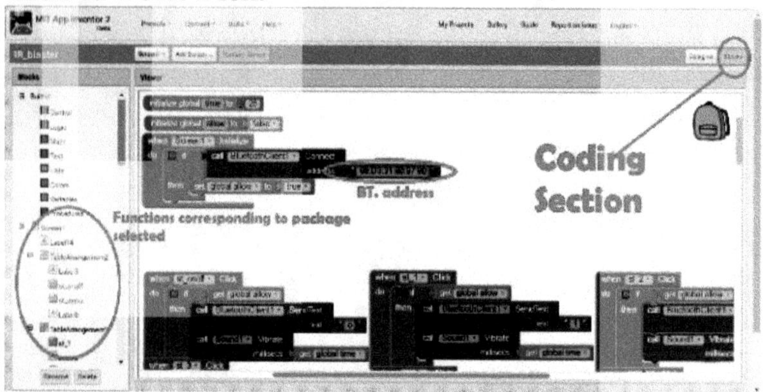

On the left half of screen you'll see each one of those bundles that you have chosen in the GUI area. The picture above shows what all segments are there in a specific bundle that you can utilize. Additionally see that Bluetooth module's location should be given in a printed arrangement.

SETP 5:

At the point when you feel that application is fit to be utilized and there are no mistakes too, click on the "Construct" tab as appeared above and select the subsequent choice. This will download your own made application, onto the PC, in ".apk" position. At that point simply move this .apk document to any Android gadget as well as snap on it to introduce.

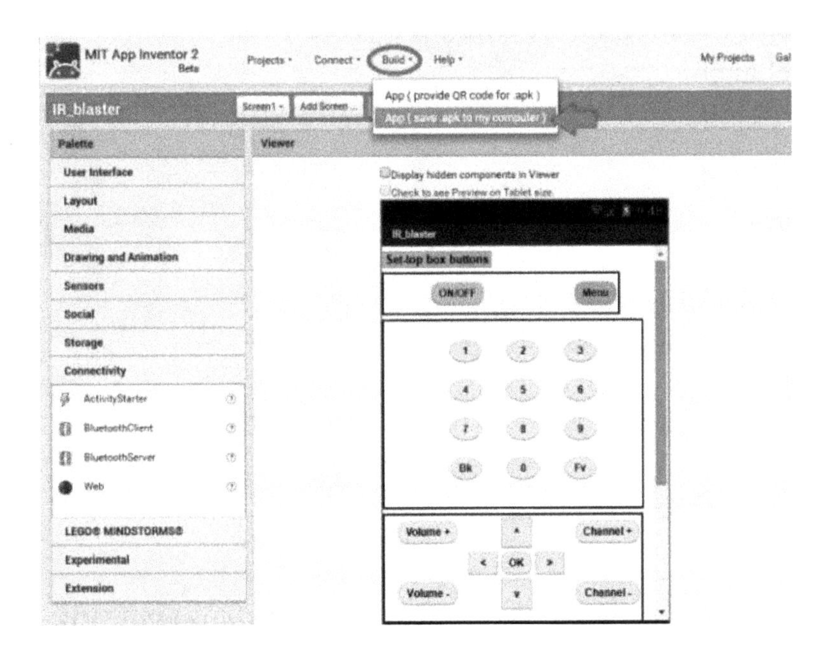

So this is the manner by which you can control any IR remote controlled gadget with your Smart telephone, you simply need to disentangle the remote of any apparatuses which you need to control with your Phone and supplant the decoded HEX code of remote fastens in the Arduino code.

Code

```
#include <IRremote.h>
#include <SoftwareSerial.h>
```

```
IRsend irsend;


SoftwareSerial mySerial(9,10); // RX, TX
char ch;
int i=0;


void setup()
{
// Serial.begin(9600);
 mySerial.begin(9600);
}
void loop()
{
 ch=mySerial.read();


  // TV buttons: ON/OFF, Mute, AV/TV, Vol+, Vol-, OK,
UP, Down, Left, Right, Menu, Exit, Return
 // Set-top box buttons: ON/OFF, 1, 2, 3, 4, 5, 6, 7, 8, 9, 0,
V+, V-, Ch+, Ch-, Back, OK, Up, Down, Left, Right, Menu,
Fav
 switch(ch)
 {
  //TV
  case 'O':
  irsend.sendNEC(0x2FD48B7, 32);
  break;
  case 'X':
  irsend.sendNEC(0x2FD08F7, 32);
```

```
break;
case 'A':
irsend.sendNEC(0x2FD28D7, 32);
break;
case '=':
irsend.sendNEC(0x2FD58A7, 32);
break;
case '-':
irsend.sendNEC(0x2FD7887, 32);
break;
case 'K':
irsend.sendNEC(0x2FD847B, 32);
break;
case 'U':
irsend.sendNEC(0x2FD9867, 32);
break;
case 'D':
irsend.sendNEC(0x2FDB847, 32);
break;
case 'L':
irsend.sendNEC(0x2FD42BD, 32);
break;
case 'R':
irsend.sendNEC(0x2FD02FD, 32);
break;
case 'M':
irsend.sendNEC(0x2FDDA25, 32);
break;
case 'E':
```

```
irsend.sendNEC(0x2FDC23D, 32);
break;
case 'B':
irsend.sendNEC(0x2FD26D9, 32);
break;


  //Set-top box
case 'o':
irsend.sendNEC(0x80BF3BC4, 32);
break;
case '1':
irsend.sendNEC(0x80BF49B6, 32);
break;
case '2':
irsend.sendNEC(0x80BFC936, 32);
break;
case '3':
irsend.sendNEC(0x80BF33CC, 32);
break;
case '4':
irsend.sendNEC(0x80BF718E, 32);
break;
case '5':
irsend.sendNEC(0x80BFF10E, 32);
break;
case '6':
irsend.sendNEC(0x80BF13EC, 32);
break;
```

```
case '7':
irsend.sendNEC(0x80BF51AE, 32);
break;
case '8':
irsend.sendNEC(0x80BFD12E, 32);
break;
case '9':
irsend.sendNEC(0x80BF23DC, 32);
break;
case '0':
irsend.sendNEC(0x80BFE11E, 32);
break;
case '+':
irsend.sendNEC(0x80BFBB44, 32);
break;
case '_':
irsend.sendNEC(0x80BF31CE, 32);
break;
case 'C':
irsend.sendNEC(0x80BF19E6, 32);
break;
case 'c':
irsend.sendNEC(0x80BFE916, 32);
break;
case 'b':
irsend.sendNEC(0x80BF43BC, 32);
break;
case 'k':
irsend.sendNEC(0x80BF738C, 32);
```

```
  break;
  case 'u':
  irsend.sendNEC(0x80BF53AC, 32);
  break;
  case 'd':
  irsend.sendNEC(0x80BF4BB4, 32);
  break;
  case 'l':
  irsend.sendNEC(0x80BF9966, 32);
  break;
  case 'r':
  irsend.sendNEC(0x80BF837C, 32);
  break;
  case 'm':
  irsend.sendNEC(0x80BF11EE, 32);
  break;
  case 'f':
  irsend.sendNEC(0x80BFA35C, 32);
  break;
 }
}
```

THANK YOU !!!

Printed in the USA
CPSIA information can be obtained
at www.ICGtesting.com
LVHW021636061123
763195LV00004B/514

9 781707 573813